A Fine Introduction to Battle:

Hood's Texas Brigade at The Battle of Eltham's Landing, May 7, 1862

Other work by Joseph L. Owen:

Texans at Antietam: A Terrible Clash of Arms, September 16-17, 1862 (2017)
Texans at Gettysburg: Blood and Glory with Hood's Texas Brigade (2016)
Lone Star Valor: Texans of the Blue & Gray at Gettysburg (2019)

A Fine Introduction to Battle:

Hood's Texas Brigade at The Battle of Eltham's Landing, May 7, 1862

Joseph L. Owen

Publisher's Cataloging-in-Publication Data
provided by Parlew Associates

Names: Owen, Joseph L.
Title: A fine introduction to battle : Hood's Texas Brigade at the Battle of Eltham's Landing, May 7, 1862 / Joseph L. Owen.
Description: Burlington, NC : Fox Run Publishing, 2021. | Includes bibliographic references and index. | Illus., 41 photos, 1 map.
Identifiers: LCCN 2021939936 | ISBN 9781945602191 (hardcover) | ISBN 9781945602207 (pbk.)
Subjects: LCSH: Hood, John Bell -- 1831-1879. | Eltham's Landing, Battle of, Va., 1862. | Confederate States of America. -- Army. -- Texas Brigade. | Texas -- History -- Civil War, 1861-1865 -- Regimental histories. | United States -- History -- Civil War, 1861-1865 -- Campaigns. | BISAC: HISTORY / Military / United States. | HISTORY / United States / Civil War Period (1850-1877). | HISTORY / United States / State & Local / Southwest (AZ, NM, OK, TX).
Classification: LCC E473.635.O94 2021 | DDC 973.7 O--dc23
LC record available at https://lccn.loc.gov/2021939936

Cover design by Sandra Miller Linhart

Cover image *The War for the Union 1862–A Bayonet Charge*, from *Harper's Weekly*, July 12, 1862 by Winslow Homer courtesy of Library of Congress.

Published by
Fox Run Publishing LLC
2779 South Church Street, #305
Burlington, NC 27215
http://www.foxrunpub.com/

Dedicated to Thomas Linward "T. L." Owen (June 30, 1903 – April 7, 1989,) a true Christian gentleman and a wonderful grandfather.

Table of Contents

List of Images

Proclaim ye this among the Gentiles; Prepare war, wake up the mighty men, let all the men of war draw near; let them come up.

- Joel 3:9 KJV.

"At Freedom's cry they left their all,
And flew to old Virginia's aid;
Where, pressing onward at her call,
West Point was stormed by Hood's Brigade."

- C. C. Chaplin

Foreword

We've all heard the dismissive remark, "That's old news." What we see, hear, or read today becomes old news tomorrow, but in the more distant future the same old news becomes the history that we are taught and learn.

Much of historical literature relies too heavily on secondary sources, rumors, and myths, and in many cases, scholars and authors provide their own interpretations of events, allowing personal bias, opinion, and errors of both commission and omission to appear in the guise of fact. The gold standard of source material for researchers are eyewitness accounts, and those are found in diaries, journals, and letters. Although such documentation doesn't guarantee 100% factual accuracy in all instances, it's the best that later researchers have.

Another source for researchers is newspapers, and other such publications. Unfortunately, they have been notoriously inaccurate from the time Gutenberg invented the printing press, up to today. Mark Twain in fact quipped, "If you don't read the newspaper, you are uniformed. If you do read the newspaper, you are misinformed." However, embedded in the often-dubious content of old newspapers can sometimes be found gems in the form of published or excerpted personal letters, interviews with eyewitnesses, or reports from correspondents who were actually present at events.

A Fine Introduction to Battle: Hood's Texas Brigade at Eltham's Landing, May 7, 1862 is the latest in a series of volumes on the battles of the legendary Hood's Texas Brigade of the Confederate Army of Northern Virginia. This book was preceded by two similar works on the more famous battles of Gettysburg and Antietam; all three the fruits of monumental research and headache-inducing transcription efforts by Joe Owen. The books are compendiums of newspaper articles and veterans' letters on the renowned brigade's various battles and campaigns, the first of which occurred at Eltham's Landing, Virginia, during McClellan's 1862 Peninsula Campaign.

The brief but important battle has correctly been described as the "baptism of fire" of the Texas Brigade and its young commander John Bell Hood, both of whom would attain prominence in American and Civil War history. The following pages contain many important firsthand accounts of the roughhewn Texans' earliest days in the faraway lands of the Virginia Tidewater, near the cradle of the old American republic, and the capitol of their new Confederate nation. The rawboned westerners were joined at the time with Wade Hampton's South Carolinians and the 18th Georgia Infantry—regiments that would earn their own places of glory in Civil War history.

Stephen M. "Sam" Hood

Author of *John Bell Hood: The Rise, Fall, and Resurrection of a Confederate General* and *The Lost Papers of Confederate General John Bell Hood.*

Acknowledgements

I would like to acknowledge, with deep appreciation, the tremendous help of William A. "Bill" Palmer for his assistance, advice and knowledge of the battle of Eltham's Landing. Bill was a tremendous resource of information about the battle, and contributed the modern-day photographs of the area around Eltham's Landing which shows the location of Hood's Texas Brigade during the battle. Also, my heartfelt thanks and gratitude to John Versluis and his fine staff at the Texas Heritage Museum-Historical Research Center. John and his staff's assistance and support was invaluable.

My sincere appreciation and admiration to Keith Jones and Eric Wittenberg whose enthusiasm and support in *A Fine Introduction to Battle: Hood's Texas Brigade at Eltham's Landing, May 7, 1862*, made this book possible.

My sincere gratitude also goes to Stephen "Sam" Hood, a descendent of General John Bell Hood whose two outstanding books, John Bell Hood: The Rise, Fall, and Resurrection of a Confederate General and The Lost Papers of Confederate General John Bell Hood helped me understand the mindset of General Hood in his first battle as Commanding General of the Texas Brigade. Sam also generously wrote the outstanding foreword to *A Fine Introduction to Battle: The Texas Brigade at Eltham's Landing, May 7, 1862*. My appreciation goes to the Library of Congress, the Virginia Museum of History & Culture, Tarleton State University Law Library, the Texas Heritage Museum and Historical Research Center and the Texas Preservation Board, The State of Texas Archives, Rick Featherston, The House Divided Project at Dickenson College – John Osborne, Elisa Vorwoork-Wood. Adam's County Texas Historical Society. I also would like to send my sincere appreciation to Mark Lemon for his generosity for allowing the use of his great-great grandfather's photograph and letter of Captain James Lyle Lemon of the 18th Georgia Infantry.

Introduction

An estimated 50,000 to 60,000 Texans served in the Confederate armed forces. In 1861, around 5,000 were sworn in what was designated as the 1st, 4th and 5th Texas Infantry Regiments. The three regiments were assigned to the Army of Northern Virginia (ANV) and were known collectively as the "Texas Brigade." The three regiments were the only ones from Texas to serve in the eastern theater of the Civil War. After arriving in Richmond, the 18th Georgia Infantry was brigaded with the Texans until after the Battle of Antietam (Sharpsburg) in 1862.

The first commanding General of the Texas Brigade was the flamboyant, hard-drinking, and ardent secessionist, Brigadier General Louis T. Wigfall. He commanded the Texas Brigade until elected as the Texas Representative in the newly formed Confederate Congress. Soon afterwards, command of the Texas Brigade went to Colonel John Bell Hood, who served as the commanding officer of the 4th Texas Infantry. He was soon promoted to brigadier general. In March 1862, Brigadier General Hood formally took command of the Texas Brigade. Afterwards, the brigade was known as "Hood's Texas Brigade," though Hood led the brigade for less than a year.

General John Bell Hood was a native Kentuckian and a West Point graduate of 1853, in his class were two future U.S. Generals, James E. McPherson and John M. Scofield, whom he faced in battle during the war. Hood was commissioned as a brevet second lieutenant in the 4th U.S. Infantry and was stationed in California. Later, he was transferred to the famed 2nd U. S. Cavalry in Texas commanded by Colonel Albert Sidney Johnston and Lieutenant Colonel Robert E. Lee. Hood first experienced battle in the rugged Texas frontier as he fought the Comanches while leading a reconnaissance patrol at Devil's River on July 20, 1857. During the battle he received a painful wound when shot through the left hand by an arrow. He was later promoted to first lieutenant in August 1858. After the firing of Fort Sumter, he resigned his commission in the U. S. Army. Frustrated that his native Kentucky was remaining neutral, he decided to fight under his adopted State, Texas. In battle Hood's usually stoic demeanor came alive. Major Charles E. Venerable said of Hood, "in the hottest of the fight the man (Hood) was transfigured. The fierce look of Hood's eyes I can never forget."[1]

1. Chesnut, Mary, *Diary of Mary Chesnut*, Fairfax, VA: Appleton & Company, 1905, 230.

After arriving in Richmond, regiments of the Texas Brigade held elections to select company officers and senior enlisted men. Many non-Texans were assigned to the brigade; however, due to the Texans displeasure of having a non-Texan lead them, the officers were soon transferred. The officers from other Southern states who remained had to earn the respect of the Texans. Colonel James J. Archer, who led the 5th Texas and Lieutenant Colonel Bradfute Warwick of Richmond, who led a company in the 4th Texas, endured the taunts and constant hazing and was finally accepted and even admired for their "tough demeanors."

The combination of Brigadier General Hood, and the Texans was more than an ordinary command, it was a great fighting machine; The combination of the frontier battle tested Hood and the excellent marksmanship of the rugged Texans—also experienced in fighting the Comanche, Kiowa, and Apache, resulted in a unique, hard-fighting brigade. These patterns of fierce charging and fighting the Union Army, during the four years of the Civil War came at a tragic price. By the end of the war, of the estimated 5,353 men who enlisted in the regiments - three Texas and one Arkansas - in 1861, only 617 remained to surrender on April 9, 1865, at Appomattox Court House. In the fall and winter of 1861-1862, the brigade was positioned along the Potomac River in close proximity to Washington D. C., to guard against an invasion by the Union Army into Virginia. Being thousands of miles away from the Lone Star State and enduring the cold winter in Virginia was a new and unwelcome experience for the Texans. The brigade was called out numerous times to "repel a major "Yankee invasion," but these all were proven to be false alarms.[2]

To alleviate the boredom of picket duty and camp life along the Potomac, on a cold winter's evening in 1861, a dozen or more soldiers of the 1st Texas took the initiative and sneaked away from camp and crossed the Potomac to "wake up General Daniel Sickles and his troops." The Texans fired into his camp and panicked the Federals into thinking a major Confederate assault was imminent.[3] The Texans were able to sneak back into camp successfully and not a word was said.

After the Union army lost the Battle of First Bull Run (First Manassas,) command of the Federal army passed from Gen. Irving McDowell to Gen. George B. McClellan. Appointed as commander of the Federal army on November 1, 1861, by President Abraham Lincoln, McClellan boldly declared "I can do it all" and began

2. Simpson, Harold B., *Hood's Texas Brigade: Lee's Grenadier Guard*, Hillsboro, TX: Hill Jr. College Press, 1970, 77-79.
3. Polley, Joseph B., *Hood's Texas Brigade: Its Marches, Its Battles, Its Achievements*, New York, NY: The Neale Publishing Company, 1910, 16.

planning the capture of the Confederate Capital in Richmond, Virginia to bring an early end to the war.[4]

While still recovering from their loss at First Bull Run, Gen. McClellan proposed to withdraw most of his army from Manassas, and transport it by water down the Chesapeake Bay and land it at Fortress Monroe, which overlooked Hampton Roads, at the tip of the Virginia Peninsula.[5] The Confederate capital was a short ninety miles away. Federal Gunboats would secure both flanks of the army, and his troops would march up the Peninsula and capture the Confederate capital.[6] When McClellan's army began its advance in the spring of 1862, Confederate forces under General John B. Magruder guarded the lower Peninsula and was strengthened by the arrival of reinforcements. Among the reinforcements was Hood's Texas Brigade. In April 1862, the weary Texans and Georgians marched from their current position at Fredericksburg, Virginia to an area around Yorktown, Virginia.

At Yorktown, the noncommissioned officers and privates of the 4th Texas Infantry presented to General Hood a fine looking horse as "an expression of their esteem." In the presentation speech, Sergeant J. M. Bookman, who was later killed at Chickamauga on September 19, 1863, praised Hood saying, "Sir, we are proud to follow, a commander whom it is a pleasure to obey, in a word, you stand by us, and we will stand by you." Hood, sprung to his saddle, and expressed his gratitude.[7]

During the first week of May 1862, McClellan, a very cautious commander, at last set his 100,000 man army on the move, up the Peninsula towards Richmond. An equally cautious General Joseph E. Johnston, leading the Army of Northern Virginia, decided that his position was indefensible. He was probably correct in his assessment because of the vulnerability of both flanks resting on the rivers and subject to the Union gunboats patrolling the peninsula. Orders were issued for the retreat on May 3, 1862, and the Texas Brigade was assigned the position of honor – they would serve as the rear guard of the Army of Northern Virginia.[8]

Confederate forces under General Johnston, retreated up the Virginia Peninsula, and fought a rear guard action with the pursuing Army of the Potomac at Williamsburg on May 5th, slowing the pursuit. The Texas Brigade under General W. H. C. Whiting's division

4. Palmer, William A., *The Battle of Eltham's Landing, May 7, 1862* book, 2012, 7.
5. Williams, Edward B, *Hood's Texas Brigade in the Civil War*, Jefferson, NC: McFarland & Company, Inc., 2012, 53.
6. Ibid, 53.
7. Williams, *Hood's Brigade*, 54.
8. Simpson, *Grenadier Guard*, 97.

marched northwest past the hamlet of Burnt Ordinary and through the Virginia mud heading forward on what was then a main road, toward New Kent Court house. Intelligence sources reported that a large Federal amphibious force under General William B. Franklin was preparing to land at this point. Franklin's objective was to intercept and destroy the Confederate supply and artillery trains slowly going through the mud a few miles west of the river and heading for Richmond. Whiting's Division was assigned the task of preventing Franklin's regiments from completing their mission.

On May 6, the bulk of the Confederate force remained in bivouac north of the hamlet of Barhamsville. General Franklin ordered his regiments to the front after they landed, he placed his artillery in support positions on the flanks and behind the center of his troops. During this time Whiting advanced the Texas scouts and a thick skirmish line to determine the place and strength of where Franklin's troops debarked.[9]

Whiting was informed that the Federals were putting ashore both infantry and artillery in the vicinity of Eltham's Landing. He moved his division through the rain and mud towards the Pamunkey River early on the morning of May 7. Since this was the brigade's first real fight, it took a little time to get organized and begin the march. It was a muddy and sodden march that dragged on as heavy rains were pouring down. A humorous encounter occurred when General Whiting, riding up and down the column, urged the men forward. "Close up men, close up!" he said " Don't mind the little mud!" "Do you call this a little mud?," said one of the men, "S'pose you get down stranger and try it, I'll hold your horse!" Not amused, Whiting said, "Do you know who you are talking to? I'm General Whiting." "General be [damned]! Don't you reckon I know a general from a long-tongued courier?" said the soldier. Frustrated, General Whiting remained silent and rode on.[10]

The Texas Brigade led the advance with the 4th Texas and General Hood in front.[11] It was here that Hood's career and life came very close to an end. As the brigade was moving forward within the Confederate picket line, it was suddenly confronted by Union skirmishers. One of the skirmishers a corporal who was only a few yards away, aimed his Springfield point blank at General Hood. Private John Deal of the 4th Texas, raised his rifle and killed the corporal before he could fire his shot.[12] Hood's escape from death was lucky because he had ordered his soldiers not to load their rifles during the march, thinking that he

9. Ibid, 97.
10. Williams, *Hood's Brigade*, 55.
11. Simpson, *Grenadier Guard*, 98.
12. Ibid, 98.

would have time to direct their loading when the Confederate line had been reached. After escaping death, Hood dismounted and ordered the Fourth Texas to load their rifles and advance on the Federals. The Texas Brigade, advanced steadily with the bayonet, and drove Franklin's soldiers a mile and a half, through the dense woods toward the river.[13] The Federal troops moved back into the woods after the Confederates left, but made no further attempt to advance.[14]

Company B of the Fourth Texas was credited with killing three Federals six hundred yards away with Enfield rifles. W. D. Pritchard of the 1st Texas wrote "the first we ever saw killed in battle."[15] The 5th Texas during this time advanced on the right of the 4th Texas and drove the enemy skirmishers on their front. The 1st Texas advanced behind the 4th Texas and the 18th Georgia was placed in reserve.

In the meantime, Colonel Wade Hampton's South Carolina Legion, supported by the 19th Georgia Infantry, with some 2,000 soldiers, advanced on the River Road driving back the Union skirmishers they encountered.

As a second brigade followed Hood on his left, Federal troops retreated from the woods to the plain before the landing, while seeking the protective fire of Union gunboats. General Whiting directed fire against the gunboats, however, his artillery had insufficient range, and disengaged around 2 p.m. Of the three brigades in Whiting's Division that fought at Eltham's Landing, it was Hood's Texas Brigade, and in particular, the 1st Texas, commanded by Colonel A. T. Rainey, that played the most prominent part.[16] Twelve of the fifteen Texans killed and nineteen of the twenty-five wounded in the Texas Brigade at Eltham's Landing were of the 1st Texas.

William Schadt of the 1st Texas faced the grim reality of war for the first time when he learned his brother Charles was killed in the battle. His anger was fierce as he wrote to his sister, "God has nerved me with the spirit of revenge and woe be to the hated Yankee foe that ever I meet." [17]

At one stage of the battle, Companies E and G of the 4th Texas came across a pocket of about eighty Federal soldiers concealed in the underbrush and "attacked them so vigorously that they dared not run and so unnerved them that they fired volley after volley into the tree tops.[18] Sixteen Federals surrendered, the rest ran from their positions,

13. Ibid, 99.
14. http://www.civilwartroops.org/1862-05-07-elthams-landing.
15. Williams, *Hood's Brigade*, 55.
16. Simpson, *Grenadier Guard*, 99.
17. Schadt, Stuart E., *Letters Home Charles and William Schadt, 1861-1865*, book: 2015, 20.
18. Simpson, *Grenadier Guard*, 100.

however, unfortunately, they ran across the front of the 5th Texas who were lying down in line-of battle order and killed every Federal soldier who ran.[19]

The Federals abandoned their positions, one after another, until they reached the safety of the fire from their gunboats anchored on the Pamunkey. The fighting at Eltham's Landing started at 9 a.m., reached its peak at 12 p.m., and by 3 p.m., had tapered off to an occasional shell fired by the gunboats.

General Johnston was pleased with the outcome and with General Whiting's division. Considering the success his men enjoyed in executing the order "to feel the enemy gently and fall back," he humorously asked General Hood, "What would your Texans have done, sir, if I had ordered them to charge and drive back the enemy? Hood replied, "I suppose, General, they would have driven them into the river, and tried to swim out and capture the gunboats." [20]

After the battle, the Federals accused the Confederates of cruelty and barbarity. General Franklin wrote in his official report:

> **"Acting Brigade Surgeon Oakley saw one of our dead who had his throat cut by the enemy." General John Newton wrote that "besides the mangling of bodies the enemy is reported on reliable authority to have rifled the persons of the wounded and of all articles of value and to have taken portions of their clothing."[21]**

Northern Newspapers also accused the Confederates of barbarity and cruelty. The Vermont Journal newspaper published a letter from a "Vermont Sharpshooter," who wrote:

> **"after the battle at West Point many of our dead were found with their throats cut. They had wounds which showed that the rebels had perpetrated this fiendish barbarity upon hopelessly wounded men."[22]**

If such atrocities did occur, it was not just the soldiers from Texas and Georgia who committed these acts, but soldiers from other Confederate regiments were equally as guilty.

Another newspaper incredulously claimed the Confederates were using "black troops" in battle.

> **"at the battle of West Point our troops encountered black troops and were driven by them. The federal troops succeeded in capturing one of these negroes and in his**

19. Ibid, 100.
20. http://www.civilwartroops.org/1862-05-07-elthams-landing.
21. Simpson, *Grenadier Guard*, 102
22. *Vermont Journal*, (Windsor, Vermont,) May 24, 1862.

pocket was found a commission for a lieutenant in the rebel army, signed by Jeff Davis!"[23]

Years after the battle another controversy arose when an unidentified soldier of the 14th Tennessee Infantry Regiment made the unsubstantiated claim, that it was the soldiers of the 14th Tennessee, who rescued Hood's Texas Brigade from certain defeat after the Texans and Georgians were overwhelmed by the number of the Federals they fought. The anonymous Tennessean wrote:

"when the Tennesseans arrived on the field the Texans were beaten, they were being driven back by the sheer weight of the numbers of the main army, and if the flanking movement we have alluded to had not been promptly repelled by the Tennessee troops, they (Texas Brigade) must have either beat a precipitate retreat or been captured to a man.[24]

This was the only account about the soldiers of Hood's Texas Brigade being close to retreating from the Federals or being captured "to a man," during the battle.

The Battle of Eltham's Landing was Hood's Texas Brigade first taste of battle, and they considered it a significant victory. The 1st Texas Infantry added "Eltham's Landing," on their Lone Star Battle Flag. The Texans and Georgians knew they could stand up against a significant enemy force and perform well under the duress of combat. There were fewer than 10,000 troops involved on both sides, and casualties did not exceed 300, nonetheless, it was an important victory for the Army of Northern Virginia. Hood's Texas brigade was able to contain General Franklin's forces near their landing points. Franklin had wanted to move against Johnston's flank and destroy the artillery and supply trains which were vital to the Southern Army.[25]

In his official report of the battle, Division Commander Gustavus Smith wrote, "all of the troops engaged showed the finest spirit, were under perfect control and behaved admirably. The brunt of the contest was borne by the Texans, and to them is due the largest share of the honors of the day at Eltham."

In the following letters, soldiers proudly wrote back home to family and friends how bravely the brigade fought in its first battle. Newspapers throughout Texas, Georgia and the South expressed their pride about the gallantry and aggressiveness of Hood's Texas Brigade in their first battle.

23. *St. Albans Weekly Messenger*, (St. Alban, Vermont,) February 19, 1863.
24. *Clarksville Daily Chronicle*, (Clarksville, Tennessee,) July 7, 1877.
25. Ibid, 102.

The Battle of Eltham's Landing was considered small, compared to the battles the brigade fought soon afterwards, Gaines' Mill, June 27, 1862, Second Manassas, (Second Bull Run,) August 29-30, 1862, Sharpsburg (Antietam), September 16-17, 1862, and just over a year later, Gettysburg, July 1-3, 1863. In the spring of 1862, however; Hood's Texas Brigade's first combat experience proved they were a force to be reckoned with. They fought with discipline and skill.

The following accounts of the battle have been transcribed exactly the way they were written in official reports, newspapers, letters, diary entries, and interviews. The common thread that binds all these accounts together – is pride in themselves, their fellow soldiers and Hood's Texas Brigade during the Battle of Eltham's Landing on May 7, 1862.

Prologue

Newspapers throughout the Confederacy published articles and reports about the success of the Army of Northern Virginia against the Army of the Potomac at Eltham's Landing, less than 72 hours after the battle. Results of the battle were published in Texas newspapers a few weeks later.

RICHMOND OBSERVER
MAY 9, 1862

Richmond, May 9. – An official letter from General. Johnston dated at Barhamsville, 11 a. m., Wednesday, states that the enemy was landing under the cover of their gunboats at West Point.

General Johnston states that the repulse of the enemy at Williamsburg seems to have stopped their advance in that direction altogether.

By later advices we have intelligence of the affair on Wednesday. The action is said to have occurred between New Kent Court-House and West Point.

We are reported to have driven the enemy back after they had assaulted our position three different times – the last time driving them to the cover of their gunboats and taking two hundred and fifty prisoners.

The troops engaged on our side were of Gen. Whiting's division, and were commanded by Gen. W. in person.

The rout of the enemy is said to have been complete, and to have been so closely pressed by our troops that many of the fugitives were driven into the river and drowned. – Examiner.

If credit may be given to the universal statement of the officers and soldiers who came from the Peninsula to Richmond on yesterday, the affair below Williamsburg, which Gen. Johnston's dispatch briefly qualified as "handsome," was an important and bloody battle. It was a determined attack on our army while in the act of retreating from Yorktown, by the advance of the Federal army. It ended in their repulse and defeat, and the result was the check of the enemy and the successful completion of the movement ordered by the Confederate commander. The loss to our side may be found to reach twelve

The march from Williamsburg, May 1862.

(Library of Congress)

hundred; that of the enemy was probably not less than two thousand. The Confederates took a General (wounded,) a Colonel, six other officers, and three hundred and fifty prisoners, who were yesterday brought to town by a troop of cavalry. They were Yankees of the genuine blue, runts in stature, mean and vile in physiognomy, and, as a specimen of the material to be encountered in the Peninsula, were very encouraging to Southern observers.

Little is known about a subsequent engagement which took place on Wednesday near Barham, in the neighborhood of West Point, where the enemy have landed forces for a flank movement. It seems to have been a complete repulse to them. They are said to have been twice driven under shelter of their gunboats, both times with terrible slaughter. Two hundred and forty prisoners taken in this fight were last evening close to the city, and have by this time probably taken lodgings with their brethren.

What is more encouraging than either of these combats, is the confidence in themselves and their commander now well established in the army of the Peninsula. The soldiers and officers feel sure to whip the enemy whenever the encounter comes, whether at Barham, Williamsburg, or any other point along the whole line. Their moral condition is an element of incalculable value in war.

Seriously reflecting on the approaching event, we cannot help sharing the confidence of the troops. The Confederate Government seems to be aware of the importance of the issue and the Value of the stakes at hazard in the Peninsula. Of all our generals, not one is probably better informed in the profession of arms or of better game than Johnston. There is not a man in his army that does not know that the fate of the country and his own individual freedom depends on the courage, discipline, and obedience of himself and his comrades. Whether the numbers are equal cannot be certainly known, but we are satisfied that there is one of the largest and best bodies of troops in the world, in good order, between this city and the York river – an army strong and brave enough to beat any other that could be possibly maneuvered in that piece of territory which they occupy. Here there must be a real fight; and we have yet to see Southern men go down before Yankees, where circumstances permitted an approach to equality and fairness in the terms of the duel.

If we can gain a decisive victory, destroy McClellan's organization, or even cripple it in such a way as to render the capture of Richmond an indefinite and improbable event, there will be foreign intervention and a treaty of peace in a very brief space of time – a period no longer than required for the transmission of the news to Europe. If the engagement should not prove favorable to us, then there must be another, and another, and another, while a single regiment remains, before evacuating Richmond; for we repeat, things have come to the pass when such an event would be disastrous in the last degree, and would be the beginning of a struggle for which even the most sanguine could not promise a conclusion.

EVENING BULLETIN
MAY 9, 1862

Letters from the Peninsula.

At a late hour last night dispatches addressed to Gen. Lee reached here by the hands of a courier.

A general action took place yesterday, the enemy, with the bulk of his strength, having engaged our lines at a place called Barhamsville, in New Kent County, about eighteen miles above Williamsburg, and thirty-three miles from this city.

The courier left at 12 o'clock, at which time the action had become general. The fighting commenced in the morning with heavy skirmishing. The statement of the courier is that, in the morning's fight, we had repulsed the enemy four times.

At a later hour positive information was received that the enemy were landing immense forces from their gunboats and transports at Bartow's Mills.

The demonstration of the enemy appears to imply a flank movement with their transports.

SPIRIT OF THE AGE
MAY 12, 1862

An official letter from Gen. Johnston, dated Barhamsville, May 7, 11 o'clock, A. M. says the Feds were landing at West Point, under cover of their gunboats; and that the repulse of the enemy at Williamsburg, seems to have stopped their progress in that direction altogether. He says nothing of the extent of the casualties on either side.

The action of the 7th is said to have occurred between New Kent Court House and West Point. We are reported to have driven the enemy back after they had assaulted our position three times – the last time driving them to the cover of their gunboats and taking two hundred and fifty prisoners. The troops engaged on our side were of Gen. Whiting's division, in which there are several of our regiments – so that North Carolina bore a gallant part in both these important engagements.

The rout of the enemy is said to have been complete, and to have been so utterly pressed by our troops that many of them were driven into the river and drowned.

Our officers and soldiers are in the highest spirits and confident of success. This fact alone is worth fifty thousand men. Gen. Johnston is superintending every movement and commanding in person.

NATCHEZ WEEKLY COURIER
MAY 14, 1862

From Richmond.

Richmond. May 8....An official letter from Gen. Johnston, dated Barhamsville, 11 o'clock, A. M. yesterday, states that the enemy was landing under cover of their gunboats, near West Point.

No mention is made of the immensity of the engagement, but on the contrary the tenor of the letter indicates that Gen. Johnston did not expect a conflict with the enemy.

He states that the repulse of the enemy at Williamsburg seems to have stopped their advance in that direction altogether.

Prisoners taken on Monday are principally from Hinzellman's Division and part from Sumner's.

Nothing is said of the extent of the casualties on either side.

FAYETTEVILLE WEEKLY OBSERVER
MAY 26, 1862

The Fight Near Barhamsville

It was stated, and then positively denied at Richmond, that a severe encounter had taken place at Barhamsville, near West Point, on the Peninsula, two days after the battle at Williamsburg. Recently we have both Confederate and Yankee accounts of it. Why it was denied we cannot tell, as it was evidently a Confederate success. *The Richmond Whig* has been favored by one engaged in it with an account, from which we extract - premising, that the 6th North Carolina Regiment, Col. Pender's though not mentioned in this Texan account of the engagement, was in it, and has the credit, by one of high position and undoubted qualifications to judge, of having done some of the best fighting in the war. Riley's battery, which is incidentally mentioned, is also from North Carolina, and is stated to have done much of the work. Gen. Whiting of Mississippi (and not a native of Massachusetts as was once said of him,) commanded in person:

While the main column, with the 4th Texas in front, and Gen. Hood and staff at its head, were marching along the road, the General and staff were fired upon by a party of Yankees, lying in ambush. Nobody was hurt. The General waving his hat, the brigade immediately closed up, and the 4th Texas was formed into line of battle. Riley's Battery, supported by the 18th Georgia Regiment was then left on the hill and the 4th and 1st Texas Regiments pursued the march. After gaining the woods, which had to be done by marching through an old field, the skirmishers found the enemy and engaged them in the woods, driving them back steadily. They came upon any quantity of knapsacks, haversacks, &c, scattered through the woods, but nothing indicating where the main force of the enemy lay. In the meanwhile, the 1st Texas came upon them in large force, and being fired upon, were immediately ordered to charge. They did charge them gallantly, and in a few volleys of musketry, sent them scattering through the woods to their gunboats, in close proximity. Promiscuous firing kept up for a while, everywhere a blue jacket offered, which was but for a short time. After the engagement, we found on the field some 250 killed and wounded Yankees, together with 42 prisoners. Hampton's Legion, which was also on the field, but not engaged, (our informant thinks,) picked up some 82 more prisoners.

Our loss was ten killed and twenty-one wounded, as appears from the Surgeon's report of casualties. Among the killed were Lieut. Col. Black, of the 1st Texas, and Capt. Decatur. Our wounded were all brought off the field by us, as were the enemy's wounded, and all came to Richmond together.

This little skirmish, which was not enough to give our Texas boys an appetite for breakfast, has been magnified by McClellan into a battle, and he has reported that we had a force of 30,000 troops on the field while he had 20,000. The prisoners taken represent the 95th Pennsylvania, 31st and 34th New York, and 1st California as in the engagement, and say that there were fifteen regiments posted in the woods near the scene of encounter, but who were, it seems, afraid to come out.[1]

---

1. Militarymuseum.org/32dNY.html. Colonel Roderic Mattheson, commanding officer of the 1st California Infantry, was a native of New York, and lived in California. He recruited a regiment, of men from the West Coast. When this original First California was formed, it was not credited to any state, but was treated as Regular Army organization. Matheson's First California was officially known as the 32nd New York Infantry. The regiment took part in nine major battles and campaigns. The regiment was mustered out on June 9, 1863, expiration of term. The men who enlisted for three years were transferred to the 121st New York Infantry.

HOUSTON TELEGRAPH
MAY 27, 1862

The Battles In Virginia.

We have the following account of the conflicts on the bank of the York River from the *Richmond Dispatch* from the 8th.

Another battle took place yesterday, near Barhamsville, in the county of New Kent, but with what result is not known, as the courier who brought the intelligence to this city left at 12 o'clock. The enemy landed their forces from gunboats, (twenty-four in number,) at or near West Point.

The number engaged on either side is not known, but the enemy was supposed to be very large. A general engagement of the two armies is expected. The loss on both sides in the fight yesterday was very heavy, ours believed to be not less than 1,000 up to that hour been driven back three times to within range of their gunboats.

We have conflicting reports of the fight at Barhamsville yesterday, and prefer to wait for an official statement before giving publicity to rumors.

DALLAS DAILY HERALD
MAY 31, 1862

From the Houston Telegraph, 23d, inst.

The Reported Evacuation of Norfolk Confirmed.
The Virginia Blown Up.
Heavy Skirmishing

By the Central train yesterday we received the following letter, and Shreveport papers of the 15th and 17th, from which we glean the following items of interest:

Nacogdoches, May 18th, 1862.

Mr. H. H. Cushing - Dear Sir: Knowing that the news must reach you through a very uncertain channels, I take for granted that you are glad to obtain for the public in all such information so may be interesting that comes from any quarter, and I therefore drop you a few lines to give you what we have just received, which seems to come more directly and is altogether more satisfactory than any which we have lately had.

I will now give it to you as we get it, and you can take it for what it is worth. The Postmaster at Alexandria writes to our P. M. that Capt. Wise, who had just reached there from Norfolk (and who engaged in the battle of which he speaks) and who can be relied on implicitly, says that a battle was recently fought on the Peninsula; results, 5,000 of the enemy killed and 8,000 taken prisoners, with comparatively slight loss on our side.

That the Merrimac (Virginia) was blown up to prevent her falling into the hands of the enemy. That Norfolk was evacuated by our troops who fled back to join the main body of Johnston's army.

DALLAS DAILY HERALD
JUNE 21, 1862

At last accounts, the main Confederate army in Virginia, under Lee, Johnston and President Davis, was at and near Richmond, holding McClellan in check and occasionally pitching in and thrashing him.

By a powerful and well executed strategy, Stonewall Jackson has succeeded in flanking McClellan's army, most unmercifully thrashed Banks, and was in Maryland, receiving very large additions to his strength, and in all probability, ere this, has succeeded in taking Baltimore and Washington City - seriously threatening Philadelphia and other eastern cities. McClellan cannot weaken his force to oppose him for fear of our grand army at Richmond, and Stonewall has everything his own way.

WEEKLY TELEGRAPH
JULY 16, 1862

The exploit of the 8 Texians at the farm-house in Virginia, where although surprised, they beat off an attacking party of 50, and made them flee in a rout which it was found necessary to conceal by the most enormous lying on the part not only of the participants, but of the General in command, to prevent a demoralizing effect, is one of the brightest incidents of the war.

The 1st, 4th and 5th regiments, during all the time from October, '61, to May,'62, had no engagements other than skirmishing, until the battle of Eltham where they put to rout five times their number, and fought so hard that the enemy, to excuse their discomfiture, declared there were 30,000 of our men and but 11,000 of theirs. The Texas brigade and one regiment of South Carolinians were all, and embraced less than 3,000 men all told. The encomiums of Gen. Johnston, Gen. Smith, and other distinguished chieftains of that army on our brigade, prove the truth of its reported valor, and are exceedingly grateful to their friends at home. The brigade is now where it will be heard from again.

CHAPTER 1

1st Texas Infantry Regiment

Battle flag of the 1st Texas Infantry.

(State of Texas Archives)

List of Companies and First Company Officers of the 1st Texas Infantry:

Company A: (Marion Rifles) Marion County – H. H. Black
Company B: (Livingston Guards) Livingston County – D. D. Moore
Company C: (Palmer Guards) Harris County – A. G. Dickinson
Company D: (Star Rifles) Marion County – A. G. Clopton
Company E: (Marshall Guards) Harrison County – F. S. Bass

Company F: (Woodville Rifles) Tyler County – P. A. Work
Company G: (Reagan Guards) Anderson County – J. R. Woodward
Company H: (Texas Guards) Anderson County – A. T. Rainey
Company I: (Crockett Southrons) Houston County – E. Currie
Company K: (Texas Invincible) San Augustine County – B. F. Benton
Company L: (Lone Star Rifles) Galveston County – A. C. McKeen
Company M: (Sumter Light Infantry) – Trinity County – H. Ballenger

The 1st Texas Infantry Regiment consisted of men mostly from East Texas counties,[1] and was the first regiment from Texas to arrive in the Confederate capital of Richmond, Virginia in August of 1861, where the Lone Star flag of Texas was presented to the 1st Texas by Lula Wigfall, daughter of the regiment's first colonel, Louis T. Wigfall.[2]

Colonel Alexis T. "A. T." Rainey, Commanding Officer of the 1st Texas Infantry. One of the signers of the Texas Secession Document. He was seriously wounded at the Battle of Gaines' Mill on June 27, 1862. Sent back to Texas on furlough and was medically discharged from the 1st Texas Infantry.

Col. A. T. Rainey:

"I have the best set of officers and men ever banded together. The Confederate Army is fighting for those noble patriots whose souls beat in unison with ours and whole hearts swell fighting for those noble patriots whose souls beat in unison with ours and swell with sympathy for the suffering soldier who leaves all that is dear to him on earth and who perils life itself upon the fields of carnage, for his bleeding country. Those through toil and suffering have dragged their wearied limbs to the field of battle to fight, to bleed, and to die for their country's liberty and their country's homes."[3]

---

1. G. Ottotot, "Clash in the Cornfield: The 1st Texas Volunteer Infantry in the Maryland Campaign," *Antietam the Maryland Campaign of 1862*, Campbell, CA: Savas Publishing Company, 1997, 73.
2. *First Texas Infantry, Handbook of Texas Online.* http://www.tsaonline.org/handbook/online articles/qkf13.
3. Letter, A. T. Rainey to Anne Rainey, August 28, 1861, Rainey Family Papers, Judy Ostler.

Confederate Postmaster General Reagan in a letter to Rainey's wife after the battle:

"The Texas Brigade engaged the Federal forces near York River and fought as Texians in most gallant style. Colonel Rainey is said to have led his regiment splendidly, remaining in front the whole fight...persons who were in the field spoke in unmeasured terms of his praise of his conduct."[4]

Colonel Rainey at Eltham's Landing.

Colonel A. T. Rainey, Commanding Officer of the 1st Texas Infantry.

(Texas Preservation Board)

At Eltham's Landing, Rainey's Company felt their first taste of gun fire. The First Texas Regiment, which Rainey was now in command, was being held in reserve by General Hood, when they were suddenly fired upon by Union troops. Rainey ordered his men to move back into the trees and "...fall on your knees, unfurl the Lone Star flag, aim low, and give them hell." The Union advance was stopped some 30 yards from Rainey, and after 20 or 30 minutes of close fighting, Rainey ordered a charge and with a wild Rebel yell, the First Texas leaped to their feet and drove the enemy back. They did not cease until General Hood himself, who had been leading the rest of the Brigade elsewhere on the battle field, rode out and ordered the First to half and retire, leaving the Federal troops in complete confusion.[5]

---

4. Letter, John H. Reagan to Anne Rainey, May 9, 1862, Rainey Family Papers, Judy, Ostler.
5. Judy Callaway Ostler, Col. A. T. Rainey: First Texas Regiment, non-published compilation, 2000, 2-7. She is the great-great granddaughter of Col. Rainey and compiled a history about Col. Rainey from family papers, letters and interviews of family members.

Lieutenant Colonel Harvey H. Black, 1st Texas Infantry.

(Texas Heritage Museum Historical Research Center)

RICHMOND DISPATCH
MAY 15, 1862

Texas Brigade. - The remains of Lieutenant Colonel Black and private Bush, of the Texas Brigade, were buried on Thursday last at Cedar Hill, New Kent county, Va., in the family burying ground of Dr. John Mayo. - These gallant men fell whilst nobly defending the soil of Virginia from the invasion of the Hessian hordes who are now trying to overrun the Old Dominion and enslave her patriotic people.[6]

TIMES-DEMOCRAT
SEPTEMBER 5, 1873

The Richmond Enquirer of the 31st ult., publishes a call for information as to the locality of the burial place of Lieut. Col. H. H. Black, of the First Texas Infantry, who was badly wounded in 1862, in the fight with the Federal troops near Eltham's Landing was brought to Richmond for surgical treatment and died.

---

6. Lieutenant Colonel Harvey H. Black, a native Kentuckian, moved to Jefferson, Texas in 1860. At the onset of the Civil War, he enlisted as captain in Company A, 1st Texas Infantry – also known as Captain Black's "Marion Rifles." He quickly was promoted to major in December 1861 and to lieutenant colonel in January 1862. He was a well-liked officer in the 1st Texas. During the Battle of Eltham's Landing, he was shot and instantly killed by Federal pickets. He became the first officer in Hood's Texas Brigade to be killed in battle. Jennifer Bridges, "Black, Harvey H." Handbook of Texas Online, accessed April 23, 2019, https://www.tshaonline.org/handbook/entries/black-harvey-h.

AUSTIN AMERICAN-STATESMAN
NOVEMBER 28, 1947

This Day in Texas by Curtis Bishop.

The Texans marching to war on this day in 1861 had a flag of their own, as Texans usually do. It was the banner of the First Company of Texas Confederate Volunteers and it was the Texas Lone Star on a field of blue. In song and story it has become known as the "Bonnie Blue Flag."

The Texans discarded their own pennant for the official Confederate flag after reaching Virginia, but at the battle of Eltham's Landing, Colonel A. T. Rainey ordered the ensign to unfurl the Texas flag instead. The Texans were on the retreat until the white star in the blue field was unfurled. They recovered, of course, and went on to victory.

A New Orleans showman seized upon the incident, and the flag, to introduce a song concerning it in an 1863 variety show. There were later versions of the tune, until the "Bonne Blue Flag" became the famous emblem of the Confederacy.

Major Matt Dale
1st Texas Infantry

Near Richmond, June 7, 1862

Dear Brother,

I will avail myself of the present opportunity to write you a line or two. You have, no doubt, long since heard of the fight on the 30th and 31st. Ever since that time we have been right in the faces of the enemy with them cannon's balls and shells falling all around us and this is my excuse for not writing to you any sooner. It is not worthwhile for me to attempt a detailed description of the fight. I have not the time even if I had the ability and the inclination. Suffice it to say that the Brigade to which I belong did not get fairly into the fight, although we charged the woods in almost every direction where it is supposed the enemy were and the Gatling shell and falling grape that were falling around us told that they could not be far from us. We passed through two or three Yankee camps and after tramping around from 3 ½ o'clock until dark. We lay down upon the field where we remained all night. The next morning about light we took a position in the woods about fifty yards in the rear of where we camped and remained there until late in the morning. We had an expectation of being attacked but aside from the roaring of a shell and canister at us we passed the day without interruption until evening when we were withdrawn about a mile and have camped since that time. With 24 hours exception we have been in front. We came to our present camp to rest. We have had some hard service as it rained on us a great portion of the time.

The men generally well. My own health is now tolerably good but I have contracted diarrhea at Yorktown and have had it from time to time ever since. It has reduced me somewhat but I am in hopes I have got it checked now. The Regiment to which Brother Jimmey is attached (the 14th Tennessee) was actively engaged and his company had eleven (11) killed and wounded through which he escaped without a scratch. His health is good and the Captain of his company being sick he is in command. Being as I wrote you though Jimmey told me yesterday he was well.

We all feel satisfied that a great battle will take place shortly here. Many of us will be laid low. If this should be my lot, I want you to manage my little affairs. Pay the last cent I owe with lawful interest. Divide what may be left equally between mother, brother, Isaac, little boy Matt, and your own children. Educate the children well, teach them to hate the Yankees and I will rest easy beneath the sod. While I write the cannon are booming in front of us and it may be that the fight will commence today. At any rate, it cannot be delayed many days. As I cannot think of anything else that would interest you, I will close by sending compliments to all inquiring friends and love to yourself, Sarah, and the boys.

Respectfully your brother,

Matt Dale

N. Biller, A. Johnson request me to say through you in Dr. Melberton to inform his family that he is well and that he wants him to see to his affairs whenever he can. See that his family have proper medical attention when they need it also that he will write as soon as he has an opportunity.

I forgot to say that our men on Saturday night provided themselves with a large number of blankets, haversacks, well filled with coffee, sugar, and oil cloth overcoats from the Yankee camp near which we

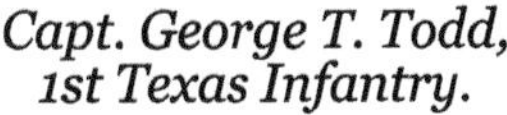

Capt. George T. Todd,
1st Texas Infantry.

(Texas Heritage Museum Historical Research Center)

camped. The Yanks left in such a hurry that they took nothing with them. They ran like the hounds they are.[7]

Captain George T. Todd enlisted as a private. He became Captain of Company A, 1st Texas Infantry. He was wounded at Antietam. After the Battle of Chickamauga he resigned as Captain of Co. A, and returned to Texas to become Adjutant of the 3rd Texas Cavalry.

Captain George. T. Todd
1st Texas Infantry

---

7. Tommy Wilson, "The Last Will of Matt Dale," in The *Junior Historian*, Volume 16, No. 6, May 1955, 6. Major Dale was killed at the Battle of Antietam (Sharpsburg,) on September 17, 1862. His Brother William had the letter probated and accepted as his last will and testament.

MARSHALL MESSENGER
JULY 16, 1897

Some Errors Corrected.
Additional Light on the Life of Col. F. S. Bass. The First Texas In More Dated, With Other Incidents of Its History by Capt. Todd
Jefferson, Tex., July 14, '97,

Ed. Marshall Messenger:

Dear Sir: I have read with interest your historical review of the life and war service of Col. F. S. Bass, the last commander of "Hood's Brigade," C.S.A. In the main the view is correct, but some errors are made which call for notice, and might mar the record if not corrected. As one of company A, 1st Texas Regiment, battle on Raccoon Ridge, Tenn. I can speak from personal knowledge. Leaving New Orleans, as you say June 1861, we went into camp, not at Lynchburg, but at the Fairgrounds in Richmond, where the regiment, not a battalion, was organized, and where we remained till after July 21, the date of the First Battle Manassas. Our regiment of 11 companies, including Company L Capt. McKee from Galveston, and over 1,300 strong were placed in a long train of box cars, just after that battle, and on the Richmond & Fredericksburg Ry. Going to the front.

Our first loss of life was occasioned by a wreck of the train, smashing into kindling a number of cars killing outright 40 of our men. We then were established in camp at Dumfries, near Quantico Creek on the Potomac, guarding masked batteries at Cock Pit Point, where we remained till February 1862. Col. Hugh McLeod had then been placed in command of the regiment, Col. Wigfall having been promoted. In February 1862 we were transferred to the Peninsula opposing McClelland's advance, and the day before the battle of Williamsburg. I think in April 1862 our division commanded by Gen. Whiting fought at the battle of Eltham's Landing, near West Point river driving back the enemy to their boats in the York river Capt. H. H. Black9 and Lieut. Harry Decatur, both of my company, and several others were killed there. Black was Lieutenant Colonel when killed. Soon after and just before the 7 pines battle our 12 months term expired, and we were reorganized for 3 years or the war, with A. T. Rainey Col. P. A. Work, Lieut Col., and Dale, Major. I was elected captain company A, and Capt. F. S. Bass became the Senior Captain of the regiment. Hood's Brigade was then organized.

You correctly outline the history of the regiment, omitting only our transfer by rail, and Whiting's entire division to the valley where we joined "Stonewall" Jackson, and under him marked a foot back by Ashland in rear of McClelland and fought seven day's from Gains Farm to Malvern Hill under Jackson.

Captain George T. Todd
1st Texas Infantry

On May 6th, or the 23rd birthday of this scribe, we met the enemy who had landed at Eltham's Landing from gun boats, and were advancing to attack Gen. Johnston in flank and rear. On 7th we fought the battle of Eltham's Landing, or West Point, driving the foe in rout and panic back to their gun boats. In this action the 1st Texas loss Lt. Col. H. H. Black and Capt. Harry Decatur and 8 men killed and wounded. Gen. Hood commanded and were much enthused over our success in our first fight. We buried Col. Black who had been the Capt. of Company "A" in a private graveyard on the hill, and the burial service of the Episcopal church was read at the grave by a lady to whom the premises belonged. He was a native of Indiana, having moved to Texas before the war, and a noble and gallant gentleman and soldier.

Dr. A. G. Clopton, formerly Capt. of Company "D," was major of the regiment in this battle. We camped some time at Baltimore Crossroads. The weather and roads were both terrible with rain and mud. Here our 12 months term of service being about to expire, May 16th, 1862, we re-enlisted "for 3 years or during the war." We reorganized by electing and field and company officers. Capt. A. T. Rainey of company "H," was elected Colonel, Capt. P. A. Work, Company "F," Lt. Col. And Capt. Matt Dale, Major. This scribe, Geo. T. Todd, was elected Captain of company "A," his old Company, and served as such the following 2 years.

We soon moved to our lines between Richmond and the Chickahominy creek, where we remained till May 31st, when under Hood, Whiting and Longstreet were ordered to attack the enemy.[8]

---

8. Todd, George T., *First Texas Regiment*, Waco, TX: Texian Press, 1964, 4-5.

2nd Lieutenant J. C. S. Thompson
1st Texas Infantry

Galveston Weekly News
July 2, 1862

Most of our readers in this vicinity will recognize the author of the description of Eltham in our paper today, as Lieut. Thompson of the "Lone Star" Rifle Company of Galveston, whom Capt. McKeen made very favorable mention in a letter we not long since published.

Battle of Eltham or West Point.

Eds. News – The enemy landed a large force at West Point with the evident intention to attempt to cut of the retreat of the army, or at least Gen. Longstreet's Division, who had given them such a large repulse at the battle of Williamsburg, which had been termed by Gen. J. E. Johnston, as a "hands on affair." The Texas brigade, then bivouacked, in hearing of their voices, the rumbling of their artillery and the "escape of steam" from their gunboats. Early the following morning Gen. J. B. Hood and staff rode up to Col. A. T. Rainey's Headquarters in the woods and remarked to him, "our front Is now clear, and no sign of an enemy ahead to impede our march, but I am going to draw in the enemy's pickets with the 5th Texas Regiment." This regiment belonged to the force ahead at West Point. All was now bustle in our regiment, preparing for the events of the day, and many questions were asked, all desiring to know if we were to be disappointed as we had been; but we did have a "brush" with the Yankees at last. Soon the 5th Texas made its appearance and passed; then came the 4th, we then knew the 1st Texas would come next. The order at last was quietly given; and we were so, armed and ready to march. We marched about a quarter of a mile and halted to throw off our knapsacks, blankets and all clothing that would prevent one from the free use of himself. We soon moved on to the scene of the action, closely followed by the 18th Georgia regiment. After a few counter marches and the (?) of guns, the 1st Texas followed in the wake of the 5th and 4th Texas, leaving the 18th Georgia regiment to support Reilly's battery upon the hill. We marched down the hill about half mile, and by the left flank, entered the woods and continued in the direction of the firing, which seem to be on the right. In advance was Hampton's Legion; the next in order were the 5th and 4th Texas regiments; then the 1st Texas wheeled to the left in order to advance in line of battle. The wheel had no more than to be finished, when Colonel Rainey was informed that the enemy was advancing, by the flank and in line of battle, forming a V in order to turn our flank, as we supposed and the prisoners captured afterwards informed us. Col. Rainey instantly faced the regiments about and right-wheeled into the road. The enemy by this time halted and smashed themselves. Two companies were ordered by Col. Rainey to advance and the regiment followed closely for forty yards and then halted. The enemy soon drove the skirmishers back, and the regiment was soon ordered to have the skirmishers come in. Then the regiment halted in the road a few moments, then fell back few yards to the ditch, and faced the enemy. Then the 3rd New York, or 1st California regiment, as it is sometimes called, fired upon our regiment in ambush, in a volley killing and wounding some of our most brave and gallant men and officers. Among those that came directly under my notice were those of our company, the Lone Star Rifles, of Galveston, who had two noble boys killed, namely, Joseph F. Brown and Charles Schadt. Smith Sims, Frank Nichols and John Coffee were wounded, the latter being wounded with a bullet in his back, which set his (?) but it is an agreeable fact that he is now up and about the streets of Richmond.

This was to us, young and unexperienced as we are, a crucial moment, who made the stoutest heart quit, but we were soon assured, when we thought of our past hardships, and all resolved to avenge our companions, this unto the death. I noticed the brothers fighting over the bodies of their dead brothers. The regiment here laid down and loaded their guns and fired promiscuously for three quarters of an hour, but the enemy's bullets went by in every direction like hail, but our 1st Texas determined to stand their ground and instantly Col. Rainey ordered the regiment to charge, which was done with the "Lone Star War Whoop," the enemy was yelling at the same time, making the scene sound as a roar, but the 1st Texas had formed in line of battle and pressed forward, by Corp. G. A. Branard, 20 feet in advance of the regiment. Branard's conduct won the applause of the regiment, and for his gallantry, he has been promoted to the rank of Color Sergeant of the regiment. It was during this charge that the brave and gallant Lieutenant Colonel Black received three mortal wounds, and was borne from the field.

The regiment formed in line of battle in the following order: the 5th Texas to the extreme right, next the 4th Texas, then the 1st Texas. The "Yanks" line of battle was (?) 21st and 3rd New York or 1st Cal, the 25th and 96th (?) in the right where they went into a swamp from which they soon retreated. We presume they attempted (?) of flanking us, but they did not succeed. We charged them this time, and the fourth charged for suppression of the field of the "York," taking the quick step out of the woods into the adjoining field, there they (?) extensive, which (?) guns of 15 regiments with their artillery under the command of Gen. Franklin, according to their account of the great battle of West Point. We moved forward, up upon the brow of a hill within a half mile of their gunboats, where we halted and rested. The gunboats kept up a large and continued bombardment, as long as we remained upon the field. The "brush" commenced at half past eleven A.M., and closed at half past three P.M. Not more than twenty minutes before, Col Rainey informed the regiment, that Gen. Smith had ordered him to hold this position at all hazard. The 1st Texas responded with "we will," then we left the field, the men feeling much chastised. Our generals knew however, that they had accomplished their designs, and frustrated Gen. McClellan's nerve here was to the slaying, where there was the enemy. Not a small arm was fired at us, when we marched by the right flank from the field, another evidence that the enemy had gloriously fled. Company C went in with a number of recruits unarmed, and the brave Capt. Decatur lost his life while arming his men with the Yankee guns picked up on the field. Many picked up guns which we found contained a ball. Whether the Yanks did it intentionally during the excitement I cannot say.

Our men were cool and contained their presence of mind throughout. We supplied ourselves with Yankee canteens, and especially with their well filled haversacks in which we found boiled ham, ground coffee, sugar, butter biscuits and soda crackers.

The enemy acknowledging the loss of 300 killed and wounded, but nothing of the missing. The 4th and 5th Texas Regiments took 44 prisoners; three were taken by Co. A. Hampton Legion 82 more making total prisoners captured 129.

The conduct and remarks of the prisoners were novel and amusing. They were very polite and admitted we had done good fighting. They said they thought we would hang them, but that in this they were agreeable disappointed. They were principally Irish and Dutch, but had been in the country some time. They were well clad in fine army blue cloth. The Pennsylvanians clothes were trimmed in red and the New Yorker's in blue.

The prisoners said that the sight of our Lone Star Flag created an unpleasant impression upon them, as they knew Texians were good "bush-wackers." They said every one of our shots seemed to take effect, and that they became disheartened at the fearful and terrible havoc created in their ranks by our well-directed fire. The loss of the Texas brigade was 40 killed and wounded – 38 of them being members of the First Texas Regiment. After the fight our brigade returned to the camps, feasting on Yankee provisions and discussing the events of the day. Thus ended the battle of Eltham: the Yankees fleeing to their gunboats as they did just before at the battle of Shiloh. Our Texas Brigade is now in the vicinity of Richmond, all in good health and spirits, and anxious to meet the enemy again, hoping to be able to catch them next time before they can reach their gunboats. The enemy's account of this battle is one tissue of falsehoods, as has been the case so many other instances.

J.C.S. Thompson.[9]

3rd Lt. Robert H. Gaston.

(Texas Heritage Museum Historical Research Center)

18 year old Robert Hugh Gaston, enlisted with older brother William in the 1st Texas Infantry. He was quickly up the ranks to 3rd Lieutenant. He was killed at the Battle of Antietam on September 17, 1862, and is buried at Washington Cemetery, Hagerstown, MD.

3rd Lieutenant Robert H. Gaston
1st Texas Infantry

May 19th, 1862

Richmond, Virginia

Dear Pa & Ma.

It is with pleasure that I improve the present opportunity of writing you a few hasty lines to let you know how we are getting along

9. J. C. S. Thompson was a member of the Lone Star Rifles and was elected 2nd Lieutenant in May 1862. He was killed at the Battle of Sharpsburg (Antietam) on September 17, 1862.

since I last wrote. I have been very anxious to write for several days knowing that you will hear many reports of our little skirmish. We had just got to camps at Yorktown & got rested when the army commenced falling back & we had to start out immediately on the march without tents. We had some rain and the roads were very muddy and you may know that there was nothing pleasant about marching with so many troops. We have been about ten days on the march & are now camped within two miles of Richmond. One of the recruits died on the march. Wm. P. Simpson of Jacksonville.

Hartwell Moore stands it finely. He has marched all the way never complaining in the least. He was in the fight & stood firm as the loudest. He said it made him feel sick when he saw some of the company shot down by his side. We frequently would go all day without anything to eat & we never get more than a few crackers a day. Frequently they would give us flour & we would have nothing to cook it on but our ramrods & frequently they would not give us time to cook it in this way. The Yankees followed us up very closely staying within two or three miles of us all the time. On the morning of the 7th we were within 2 miles of West Point on York River & we were informed that the Yankees had landed a large force of 15 to 20,000 men at that place. As they were attempting to cut off a part of our army which was several miles behind, we, of course, had to attempt to drive them back. Gen Hood & Gen Whiting were both to engage them with their brigades. Gen Hood was marching us down towards the enemy's position with skirmishers thrown out in front when we (our regiment) were attacked by a large force of the enemy (which we afterwards learned to be three regiments). As they attacked our left wing before we were expecting it we were somewhat confused & fell back about 200 yards to an old ditch where we halted & waited from them to come up. As the place was very brushy, they could not see us & supposed we had left the field. So they raised a yell & came charging at full speed. But when they came within 80 yards of us, our boys rose out of the ditch & poured such a volley into them that their ranks commenced wavering to & fro not knowing what to do. Col. Rainey seeing this, ordered us to charge them which we did promptly & they fled in the greatest confusion. We pursued them 2 or 300 yards & then halted for a few minutes when we were marched back to camps. Our boys picked up many things such as canteens, knapsacks, haversacks full of provisions etc. Billy got a fine sword. Our regiment lost about 15 men including Lt. Colonel H. H. Black. Our company lost one killed on the field, Perry Mills of Md. Prairie. 4 others have died. Lt Spencer had his right forefinger shot off & went to the hospital & we have just received word that he is dead. I suppose the wound mortified. Sam Cornwell of Kickapoo was shot through the head and died a day or two after the fight. D. J. Hill of Jacksonville was shot through the head and died the next day. W. A. Hones of Jacksonville was shot through the hip & lived three days. Fayette Martin of Jacksonville had 1 finger shot off & he is doing well. John Foster of Kickapoo was grazed by two balls, one on the arm, the other burning his eye severely, but he is able for duty. Capt. H. E. Decatur of Marshall was killed. We have heard that it has been written back that Billy was killed. But through the kindness of heaven neither of us were hurt. No other regiment except ours was engaged in the fight much. The fourth Texas had a little brush. I would be glad to write more but I have no more time & must close. Write often & we will write as often as we can. No more at present.

Your affectionate son,

Robert[10]

Color Sgt. George A. Branard became Color-Sergeant of the 1st Texas Infantry at the Battle of Eltham's Landing. Except for the Battle of Sharpsburg, he bore the colors of the 1st Texas.in every battle up to the Battle of Chickamauga when he was seriously wounded. He was awarded the Confederate Medal of Honor for actions at the Devil's Den, during the Battle of Gettysburg, July 2, 1863.[11]

Color Sgt. George A. Branard
1st Texas Infantry

San Antonio Express
March 18, 1900

Eltham's Landing.

---

10. Glover, Robert L., *Tyler to Sharpsburg: The War Letters of Robert H. and William H. Gaston Company H, First Texas Infantry Regiment, Hoods Texas Brigade*, Waco, TX: Texian Press, 1960, 7-8.
11. On July 28, 2018, Color Sergeant Branard was awarded the Confederate Medal of Honor by the Sons of Confederate Veterans for actions above and beyond the call of duty at the Devil's Den during the Battle of Gettysburg on July 2, 1863. His Confederate Medal of Honor is on permanent display at the Texas Civil War Museum in Fort Worth, Texas.

Color Sergeant George A. Branard.

(Patty Branard-Gambino)

These few remarks relate to the fight at Eltham's Landing, a battle that has not appeared in print neither have I seen it in what is called the "official reports" of the United States War Department, although I may have overlooked it in some other reports of which I have many. Now, I am stating only what I saw and am not trying to give a full detail of the fight, so no comments are needed as to my mistakes as to how others saw it, for no two persons ever saw a fight alike. Hood's Texas Brigade as usual brought up the rear of our army in falling back from Yorktown. After falling back as far as Williamsburg, we camped for the night in an old field not far from the woods. The next morning the Yankee cavalry appeared in sight when a brush between our cavalry took place. As to the results we won the fight and had some few sore heads among our buttermilk rangers. I saw one of them that knew how to handle his saber, had made a cut at and had taken a piece off the top of his head, about the size of a half a dollar. In talking with him, I told him that he never could be a whole man, as the Yank had cut his cuticle, scalp skin and all, and his hair never would grow back at that spot.

After this little rumpus everything became quiet and our brigade got orders to march towards river by the river which lead from Eltham's Landing and so we took a road leading from Williamsburg to that point. When night came we went into camp alongside the road. Rations for two days was issued, such as flour and bacon – half a pound of bacon and one pound of flour was given for two days' ration. The bacon we knew how to cook, but as our wagon with cook utensils was not to be found, many of us scratched our heads as to what to do with it. Finally one of us, wiser than the rest, skinned the bark off a tree and mixed his flour in it, and it made very good bread, through both ends open. Another one thought it might cause the bread to taste of the sap, so he used his hat, turning it inside out, and making a bowl of it. The next thing was how to cook it but that did not trouble us long. Some got stocks and others their ramrods, wrapped or rolled the dough around them and stuck the other end in the ground and cooked it as though roasting bread. Some made flat pones of the dough and laid it on a chip. Well sometime in the small hours of the night we managed to use up all the flour, letting the half done dough get cool while we slept. Our boys being honest not one of them thought of taking anything that did not belong to him. We all slept soundly without fear of losing our night's cooking. At the peep of day we fell into line of march. My regiment, First Infantry, I think, was in the rear that day, and we moved off down the road to Richmond.

About 8 o'clock or somewhere near that time, word came back that the Yankees had attacked the head of our column and killed W. D. Denny of the staff, who was killed by a sharpshooter. Our regiment was ordered up at double quick. The regiments in front were ordered to charge them. One company of the Fourth Texas was sent out as skirmishers, and drove the Yankees back fully a thousand yards, killing some. As to where the Fifth and Fourth went I did not know, but when the First arrived at the place of confusion we were ordered down to the road, and while marching down the road that led to the landing our regiment received a galling fire from the left and somewhat in our rear. The attack was so sudden that we halted not knowing at the moment what to do, but not for long, when Col. Rainey gave the order for the left wing to fall back. Then we became very badly mixed up as our regiment had gone into this fight left in front, and the order being given for the left to fall back both wings came back together when he intended the wing on the left to fall back and by facing us into the front it would throw us into the shape of a V, facing us both to the right and left, so as to meet the Yankees from either direction. We fell back some twenty feet and reformed and the word "Charge!" was given, and do or die trying. Every one of the brigade and many of the general officers of the division and corps know the result. The Federals lost 300 or more killed and wounded and we took 125 able bodied Yankee prisoners. Our loss in the First Texas was – so my report shows which I have now in my possession – 15 killed and 19 wounded.

I suppose this fight was the cause of the Yankee government putting the meanness in the army to fight us thinking they were men of fighting qualities as the correspondent of the New York Herald, in writing up the fight of Eltham's Landing – or as they called it, West Point – stated that "four regiments of negroes charged our troops furiously and drove us to the cover of our gunboats with heavy loss."

With no reflection on the toilet of Hood's Texas Brigade, I will say that we did look a sorry set, particularly when we had sat up most of the night cooking dough on sticks, and standing in the smoke trying to do it, for, as everyone knows who ever had to warm or cook by a camp fire, the smoke always blows to the side you are on, not matter how many times you change your position, and the readers of this can imagine how we looked that day, as we had no time to wash, and to wash takes water, water being hard to get or a long way to go for it.

In this we lost our Lieutenant Colonel H. H. Black, a brave man gone before the war had hardly commenced. To give a list of those killed and wounded in the regiment would take up too much space. But of my company – I. – known in Galveston by the name of the Lone Star Rifles: Killed, Jos F. Brown and Chas L. Schadt; wounded. Privates Frank Nichols, S. D. Sims and John Coffee. We had four officers and seventy-one men in action. Jos. F. Brown was in the rear rank, just behind John Coffee. Brown was in the act of firing. It was while we were kneeling down, when he was shot, and in falling over his gun went off and shot John Coffee in the right shoulder blade, plowing the flesh up and across his shoulder. Coffee knew how to nurse his wound, as he had it always sore.

An accident happened here, one in which I was an eye witness. I have told it before, in some of my writings, but as it belongs to this fight, I repeat it again. One of our companies of the First Texas had an Indian with it. We called him Ike, and Ike was always spoiling for a fight until the fight came and having lost sight of him that morning. I cannot say what part he took in it. My first sight of him that day was after the fight was over. Most of us were walking about looking at the captured prisoners. I saw him sitting behind some wounded Yanks, and he was looking at one of them and asked how he liked the fight. He said that the fight was big much fight, heap of fight, we whipped them. He looked as solemn as an owl and then remarked, "See him Yankee? He no good, me kill him."

I said: "No Ike, don't do that; he is wounded; he get well." And then I asked which Yank he meant. He pointed him out by raising his gun to shoot, when one of the guards who was in charge of this wounded squad stopped him and told him if he ever made another move like that he would put him in the guard house. Just about that time one piece of artillery was run down to rear and in rear of these wounded men and let loose a shot in reply to one from the gunboats, which had commenced to shell the woods, when Ike looked up:

"Whoop hee! Big gun on wheels shoot much, big ball hit somebody, he got hurt, Ike he no like big guns. Good-bye, Ike he goes."

With that he left, and I never saw him again, but some of his company told me that Ike got home all right and was a big man among his tribe.

I have written this as I know and saw it. Anyone else is welcome to write this fight up to the queen's taste if he chooses. I will not kick at him doing so, but do not tell me I am wrong in what I have seen and written. I have more, but as it will have to bring in self-praise I have left it out. The historians can do that.

George A. Branard.
Acting Color Bearer of First Texas at the time.

Color Sergeant George A. Branard
1st Texas Infantry

Galveston Daily News
June 26, 1903

Galveston. June 24. – To The News: The last time I saw George Branard in Texas in '61 was when the "Lone Star Rifles" marched down Tremont street to take the boat to Houston on the way to Virginia. He was bit of a dandy, a beardless youth, and about the proudest boy that ever marched off to war. All Galveston was out to see the Lone Star Rifles off. Mothers, sisters, sweethearts all vied with each other to do honor to the brave boys. George had his musket, and though placed well towards the rear on account of being the smallest man, or rather boy, in the company, he stepped as high and felt as big as the largest man there. That was the grandest day for the Lone Star Rifles. There were sore and bleeding hearts in their ranks and more of the same kind among them who had gathered to bid them good-bye and God-speed, but all put on a brave front and cheers and bouquets were the order instead of tears and sighs.

The next time I saw George was in Virginia on the morning of May 7, when I joined my company at Eltham's Landing, or as the Yankees call it West Point. A few days before at Yorktown the color sergeant of the First Texas Regiment had been killed in a skirmish some of the boys brought on with the Yankees and George Branard, happening to be one of the color guard, had taken possession of the First Texas flag.

I had scarcely delivered some messages to the boys from the home folk when we were ordered to fall in and begin our march, no one dreaming of serious trouble or what was in store for us. We had gone about two miles when the Fourth Texas ran into a body of Yankee pickets who fired on them. We were halted at once and the Fifth Texas Regiment was deployed into the woods to feel for the enemy. A moment after the Fourth Texas ran into an ambush and had several men and officers killed and wounded. The Fourth stood their ground like veterans and engaged the enemy. Our regiment, the First Texas, went to the support of the Fourth. We were all young and fresh then, but I don't think there was a single member of the First that did not realize the absolute necessity of dying right there in our tracks to save the situation. The Yankees got right upon us, when our Colonel ordered us to charge. It was a desperate thing to do – our regiment against a whole brigade, but we did it. We raised a yell and went at them with the bayonet and they broke. They formed again and poured in a terrific volley of mine-balls, and we charged them again. It was right here that George Branard won the right to keep the First Texas flag. In those days his wind was good and his legs limber and he took that flag fully a hundred yards in advance of the regiment. He jumped up on a rail fence and called on the regiment to follow him. His Colonel called attention to him and ordered the regiment to "follow their brave color bearer," which they did with a yell.

The next day the regiment was drawn up as if for dress parade and George was ordered twenty paces to the front, and then, after an eloquent address by the Colonel, he was officially appointed color bearer of the First Texas.

Eltham's Landing was the first chapter in the brilliant career of George A. Branard. A better and braver soldier never wore the Confederate uniform. At Little Round Top, at Gettysburg, he carried that flag to the farthest point reached and was nearly blown to pieces by Yankee shells for his pains. At Knoxville he got a good dose in his left arm and was placed in the hospital. Our army fell back and left him there. The surgeons announced that all of our wounded would fall in the hands of the enemy. George was badly wounded, but when he heard of this, he and two others got up and walked all the way back to Virginia. When he got back to his command he was in a sad plight and was put into the hospital at once. When he was a convalescent, a big fight took place and George was placed in charge of six ambulances. His orders were general and he exercised his own judgement. He did not know anything about ambulances; all he knew was fighting; and that day both armies were amazed to see six ambulances on the firing line. It is an actual fact, he took the whole party to the front on the firing line and came near losing himself and the whole outfit. After that they trusted no more ambulances to George's care, but gave him back his flag.

George Branard was one of the best, bravest and most gallant members of Hood's Brigade. His record as a soldier is without a blemish and I call on every survivor to bear witness to this fact.

"Lone Star Rifles."

Color Sergeant George A. Branard
1st Texas Infantry

George Branard's promotion from Fourth Corporal to Color Sergeant occurred at Eltham's Landing, Va., known to the Federals as West Point," said General Polley. "The promotion was made on May 7, 1862, and occurred in this way: Thomas Nettles of Livingston, Texas, who I believe is still living, was Color Sergeant of the First Regiment, with a guard of eight Corporals. He became impatient at the long delay in the opening of the first battle, and evidently believed he was going to be denied the opportunity of shooting at a Yankee. During the skirmish he went down into a rifle pit to get his shooting chance. While he was shooting at the Yankees, they were also shooting at him, with the result that he received a bullet in the shoulder which made it impossible for him to longer carry the flag. He transferred it temporarily to Branard. In an engagement next day, Branard" — this with a twinkle in the eye — "thinking he was going to the rear, got too far in front of the regiment to hear the command to halt and fall back. When someone shouted to him to fall back, he declared that the regiment could fall back if it wanted to, but he'd be d__ d if he fell back. The (Colonel heard the remark, and, admiring both its courage and spirit, ordered the regiment forward to form under the colors. And he then and there promoted Branard from Corporal to Color Sergeant. The Colonel was A. T. Rainey of Palestine.[12]

---

12. Chilton, Frank, *Unveiling and dedication of monument to Hood's Texas brigade on the capitol grounds at Austin, Texas, Thursday, October twenty-seven, nineteen hundred and ten, and minutes of the thirty-ninth annual reunion of Hood's Texas Brigade association held in Senate chamber at Austin, Texas, October twenty-six and twenty-seven, nineteen hundred and ten, together with a short monument and brigade association history and Confederate scrapbook*, Houston, Texas: Frank Chilton, 1911, 192.

Sergeant Malichiah Reeves, 1st Texas Infantry.

(Rick Featherston)

Malachiah Reeves was a sergeant in Company I, 1st Texas Infantry. He was captured at Mechanicsville, Maryland in 1862. After the war he became a Baptist Minister and Postmaster General of Leaguesville, Texas.

Sergeant Malachiah Reeves
1st Texas Infantry

Papa's first battle experience was at Eltham's Landing. He has the battles in which he fought recorded as he engaged in them in the little note-book in which he kept his diary. In a conversation with "old Brother Wyrick" I heard him tell of this battle. "Eltham's Landing on the York River is where I had my baptism of fire and it readied me for what was to follow." Gen. Johnston (Joseph E.) withdrew his troops from Yorktown, you know, and moved back toward Richmond. Franklin tried to intercept and hold the Confederates, so that was a battle near old Williamsburg two days before the Eltham's Landing battle. As we followed up this action our Company I with First Texas, Hood's Brigade was the "cow's tail" and our experiences were (?) to say the least! We were not to sure of where we were going, but excitement ran high. It was a nasty rainy morning on the fifth of May and we were sloshing through mud over our shoe-tops. The roads were completely obliterated by our tramping. We stopped for the night. The supply wagons finally caught up with us and we had a feed that was certainly not "fit for a king," but it was good for hungry boys such as we.

On the morning of the 7th we were headed for, we know not where, when out of the nowhere came Yankee picket bullets. As our commander rode ahead of us one of the bullets was spent for him, but a fellow comrade had loaded his gun (this he was not supposed to have done) and he let the Yank have that load. From then on for the day we were under fire and returned as heavy as we received. Our loss was not so heavy, but Col. H. H. Black of our Company I was mortally wounded. We buried him therein the deep dark pine woods. Only God knows where that grave is located!

"Brother Wyrick, do you remember the battle in the apple orchard? We lost Sapp in this battle; the only casualty. I know that our First Texas Regiment was just about the best in this fight. We drove the First California to the river. Our troops, it is said, saved the Confederate army wagon train, supplies and munitions." Always thrilled to sit and listen to these conversations, I sat "spelled-bound" through this one.[13]

3rd Corporal John N. Wilson
1st Texas Infantry

Was hit under the arm by grapeshot at Sharpsburg, but the skin was not broken. Was wounded in the face and on the hand in the Wilderness on the plank road. Was elected Third Corporal Lieutenant in May, 1862, and was promoted to Second Lieutenant, and Finally to First Lieutenant, Sept, 17, 1862 and to Captain, May 4, 1863. Was in the battles of Second Manassas, Boonsboro Gap, Sharpsburg, Fredericksburg, Norfolk, Chickamauga, Knoxville, Wilderness, Gettysburg and Appomattox C. H. and many other small battles and skirmishes. Company K of the First Texas Regiment was raised by B. F. Benton of San Augustine, Texas, and the company armed itself with shotguns and rifles and fixed a large knife on the end as a bayonet.

Our regiment first engagement was Eltham's Landing and we did not lose a man. The next was Seven Pines, where one was severely wounded, and then at Gaines Farm, where we lost our Captain and some others. As we followed the Virginia Army to Malvern Hill, fighting more or less every day for about seven days, we lost heavily, having only one commissioned officer in the company. We next met the enemy at Second Manassas, where our brigade suffered considerably, but our company only had one man wounded. Our brigade participated in the following battles: Eltham's Landing, Seven Pines, Gaines' Farm, and from there to Malvern Hill, Second

13. Eads, Leila Reeves, Malachiah Reeves, "Memoirs of the mercies of a covenant God, while traveling through the wilderness of this world of life: being the autobiography of Malachiah Reeves, originally from middle Alabama and through much of East Texas and now of Ranger, middle West Texas. Unpublished manuscript, Hood's Texas Brigade Files, Texas Heritage Museum-Historical Research Center, Appendix A. Leila Reeves Eads was the daughter of Sergeant Reeves and transcribed his memories of being in Hood's Texas Brigade as he said them.

Manassas, Antietam or Sharpsburg, Fredericksburg, Norfolk, Chickamauga, Chattanooga, Knoxville, Wilderness, Spotsylvania, Cold Harbor, Gettysburg, around Petersburg and Richmond, and was surrendered by Gen. Lee at Appomattox.[14]

4th Corporal O. T. Hanks
1st Texas Infantry

The time had now arrived when we were to evacuate Yorktown. The whole army was on the march toward Richmond. We were attacked at Williamsburg by the enemy. He was soon driven away. We proceeded on our route to Eltham's Landing on the Pamunky River where the enemy attacked us again and put up a pretty stiff fight. We drove them back to their gunboats. Some North Carolinians were in charge and were going too far and were stopped. They complained and said if it had been those ---- Texans they would have been allowed to go on.

General Johnston moved on. When he reached the Chickahominy on the side next to Richmond, which was about seven miles distant, he made a final stand. General McClellan came forward promptly and took his position. Nothing much took place, only skirmishing and cannonading.

Now, here we were facing as grand an army as was marshalled on American soil, equipped with everything for benefit and comfort. They were perfectly elated with this much success. Almost everything they possessed was marked and engraved: - On To Richmond, - which proved to be true by things that we captured when we whipped them. Now, remember all this was not accomplished in a day; it went into weeks, months, and even years.[15]

---

14. Yeary, *Boys in Gray*, 808. John N. Wilson enlisted as a private in the Lone Star Rifles, later Company "K," of the 1st Texas Infantry. He was elected to 3rd Lieutenant on May 17, 1862 and quickly was promoted up the ranks to obtain the rank of captain. He was wounded at the battles of Chickamauga on September 19, 1863, and the Wilderness on May 6, 1864. He was paroled at Appomattox on April 12, 1865.
15. Hanks, O. T., *History of Captain B. F. Benton's Company, Hood's Texas Brigade, 1861-1865*, Austin, TX: Morrison Books, 1984, 7. O. T. Hanks was a soldier of Company K of the 1st Texas Infantry. In November 1861, he was listed as sick with measles, and in a hospital in Richmond, Virginia. He was wounded in the battle of Antietam (Sharpsburg). In February 1863, he was listed as wounded in the left side in an area around Suffolk, Virginia. In April 1863, he was again listed as sick in Richmond. On October 14, 1864 he was promoted to 2nd Sergeant, and was present at Appomattox when General Robert E. Lee surrendered at Appomattox Court House.

Private Ed Buckley
1st Texas Infantry

GALVESTON DAILY NEWS
AUGUST 2, 1891

Hood's Men In Battle.

The Story Told By An Old And Interesting Letter.

The First Battle in Which the Famous Brigade Engaged - Pathetic Memories of the Times of War.

Houston, Tex, July 31 - The following is the chief portion of a letter now in the possession of Mr. George Branard of this city. It is yellow with age, tattered and torn, but is prized very highly by the holder and many of its references are to persons well known in Houston and Galveston. It is an account of the first battle of Hood's Texas brigade, afterwards one of the most distinguished brigades in either army. It was written to Dr. J L. McKeen of Galveston soon after the battle. Mr. Buckley, the writer, lives now in LaSalle county and has charge of the T. W. House land and stock interests in this county. Mr. George Branard, the gallant standard bearer, lives in this city. Louis Coleman, Jo Cramer, F. M. Poland, John Cameron and Major Burns of this city were in the fight. It was called by the Confederates the battle of Eltham's Landing, by the federals the battle of West Point. It is dated Camp of Texas Brigade

At Balt. Store, 1862. - Dear Doctor: Yours truly welcome favor reached one at Yorktown a few days prior to our evacuation of that point. The marching and fighting occasioned by our retreat from there to this place has made it impossible for me to write to you before now, and though my convonleuce for writing is very limited, I owe it as a duty to my friends in Galveston to write and give them some account of the battle of the 7th instant of the Texas brigade. At about 8:30 a. m. our brigade, composed of the First, Fourth and Fifth Texas regiments and the Eighteenth Georgia received marching orders. When we had marched about two miles on the road towards Richmond a party of Yankee pickets fired on our General Hood at a very short distance, but thank God, without hitting him. The general was marching at the head of the column when he was fired at and some of his old regiment, the

Fourth, fired at the yanks and made one of them bite the dust. That of course made us expect a fight and the general sent the Fifth Texas commanded by Colonel J. J. Archer into the woods to meet the enemy, and while moving in front with Captain Denny, his commissary, they were fired upon by Yankees in ambush, and Captain Denny was struck in the upper lip with a ball that went through his head and killed him. In the meantime General Hood took the Fourth Texas late into the field, which I believe, opened the engagement, and next came our regiment, the First and we were attacked by a Yankee brigade that was coming down and trying to cut off the Fourth. The gallant First met them at a disadvantage because we were not expecting them that soon. But as Texans, should we meet them, to do or die, and commanded by gallant officers, we commenced the deadly fire with as much coolness as the Lone Stars exhibited on the public square at Galveston when drilling to please the ladies. Never did a regiment, with greater coolness and determination, meet a foe so formidable, for our regiment did not number 600 men, while there was a brigade of the Yankees, and I believe they brought fresh troops against us after we had repulsed one brigade.

After firing a few rounds at a distance of about forty yards our gallant colonel said it would be a useless waste of life to stand before five times our number and receive their fire, and he ordered us to charge them, which was done with the "war-whoop" of the Lone Star rifles, in which the whole regiment joined. It was in this charge, while thirty steps in advance of the regiment, and shouting "Liberty or death," the braves of the brave, Lieutenant Colonel Black, fell mortally wounded. He had been twice shot without checking his advance. A third shot struck him as he was getting off his horse and went through and through his body. It was there, also that Captain Decatur was killed. He took fifty recruits into the field without guns, and had armed them all with Yankee guns before he fell. I take pleasure in stating to you that the captain of our choice, "our father," was as cool in the field as when you saw him in Galveston give the command: "Lone Star Rifles, commence firing!" He acted gallantly, and the boys stood by him with a spirit that did honor to the Lone Star state. The First Texas was tried in that fight and won the admiration and applause of the division and the whole army. The Fourth and Fifth were of course, behind us. But we happened to have the hottest of the fighting. Our loss was: First Texas: 31 killed and wounded. (Joseph L. Brown and Charles Schardt were killed and Smith Sims, Frank Nichols and – Coffee wounded, all of Galveston): Fourth Texas, 3 or 4 wounded: Fifth Texas, 3 killed and three wounded; the Sixteenth Georgia of Colonel Hampton's brigade was with us and got 1 or 2 killed and – wounded. Our regiment had the hardest fighting, and if we had not charged them with fixed bayonets they would have shot us all to pieces. The Yankees threw away their haversacks, canteens and guns when we charged them. Their haversacks were all filled and we had coffee and sugar mixed and buttered bread, luxurious that we had almost forgotten the existence of. Some of our men got Yankee guns and I did; but I gave mine away, as I already had a good gun. We took forty-four prisoners, and would have taken more but they outran us. We chased them until the shells that were bursting right in front of us from the Yankee gunboats at West Point, the terminus of the Richmond and Yorktown railroad. The Yankee loss in killed and wounded was 700 or 800. The Eighteenth Georgia was not in the engagement. It was held as a reserve. Our general J. B. Hood, is a native of Kentucky and an officer of the old United States army, and served on the Texas frontier. He is a brave officer and has the perfect confidence of his men, and he is assured now that they will follow him until the last man falls.

George Branard of our company carried the Lone Star flag in the fight and won the admiration of the officers and every man in the regiment. He is not the color bearer of the regiment, but the colonel says he has won the right to be and shall be hereafter. You must excuse this scrawl - I am writing on the ground and in a hurry, as I am ordered to start a half hour to Chickahominy creek, nine miles from here, where our wagons are stationed. See Sid Sherman and tell him I received his letter today and will write him as soon as possible. We are all well and in good spirits and confident of success. Remember me to inquiring friends. Write soon, dear doctor, and believe me, your sincere friend.

Ed Buckley[16]

---

16. Private Ed Buckley enlisted as a private in the "Lone Star Rifles," later Company "K" of the 1st Texas Infantry. He served as Aide-de-Camp to General Hood at the battle of Gaines' Mill on June 27, 1862. Buckley was appointed as the Texas Brigade Quartermaster in July 1862, and then to the Division Quartermaster Clerk in February 1863 He was assigned as Clerk for a Major George, the Quartermaster for Hood's Corps in the Army of the Tennessee in July 1864.

Private J. J. Hale, 1st Texas Infantry.

(Texas Heritage Museum Historical Research Center)

Private J. J. Hale
1st Texas Infantry

TEMPLE DAILY TELEGRAM
JUNE 25, 1913

The subject of this sketch was pointed out to me by Uncle Billie Stevens, because they were both members of that Hood's Texas Brigade that you read about. He was born in San Augustine County, Texas, January 19, 1840, and he joined the army in 1861 and went through the war. He was a member of Co. K, First Texas, Wigfall's brigade at first, but commanded by John B. Hood beginning in the spring of 1862. There were three Texas regiments - 1st, 4th and 5th. Mr. Hale's first battle experience was at Yorktown and he was among those who volunteered to hold dam No. 1. There he fired his first shot at a Yankee and killed him. The brigade was first in action at Eltham's Landing, where the 1st Texas was most involved. They were armed with shotguns and squirrel rifles, but they won the fight and captured enough guns and ammunition to fix them up pretty well. At Seven Pines, which was not more than ten miles from Richmond, the Confederate capital, they had a fight and at Gaines' Mill they got in the rear of McClellan's army and on the 26th and 27th of June they won the victory which Hood's Brigade is now celebrating in Temple. They won a victory at Cold Harbor and a new one every day until they got to Malvern Hill where Hood's Brigade stormed a cannon mounted hill and didn't take it. At the second battle of Manassas they were charged by the New York Zouaves, whom they obliterated. He was in the fight at Fredericksburg, in Pennsylvania at Gettysburg, and at the Wilderness where Gen. Robert E. Lee came to the rear of the company and talked with the captain. He was at Spotsylvania Court House, Cold Harbor and Petersburg. He was on Darbytown Road when they were charged by negro soldiers. On the seven mile line Hood's brigade was attacked by an army corps which they captured including eight stands of colors. Mr. Hale was the last color sergeant of the First Texas regiment.

Private J. M. Leach
1st Texas Infantry

The Eagle
July 4, 1925

J. M. Leach, aged 83, from San Antonio, was born in Pike County, Alabama on May 6, 1842. He enlisted in Galveston, leaving there the first day of August 1861 for Virginia. He had come to Texas in 1849 and his people settle first about 20 miles from Rosenberg at the San Bernard river. Mr. Leach was 18 years old when he enlisted. The war came on in time to spoil his career as a newspaperman, he being engaged at the time as a printer's devil on the Christian Advocate at Galveston. His company was "L" in the First Texas, Hood's Brigade, and joined Lee's Army of Virginia under General Joseph E. Johnston. They joined Lee's army at Manassas Junction and got there just 26 days after the first battle of Manassas and the field of the battle had not yet been cleared.

"We marched from Galveston to Beaumont," said Mr. Leach, "and then, when irked by the delay in getting to the battle front, the whole company of 137, confiscated a boat on the bay and set out for New Orleans. It was a small vessel. The private owner was angry at our action and grounded the vessel on a shell reef, with nothing to eat, we all got out on the reef and took hold of a rope at high tide and pulled the vessel off the reef. When we got off the vessel we made a rush for two little stores and scooped up everything that was in them. "We then tramped to New Iberia, got on a boat and crossed the stream and went to New Orleans. We stayed there about two months and in November they moved us over the Potomac. In January we pulled out for Yorktown, down on the peninsula, between the James and York rivers.

Was Never Scratched

"We thought for a time we were going to see a dual between Colonel A. H. Belo and a little Englishman. A dispute had arisen between the two men and a Englishman, a little fellow who always wore a long, black robe, wherever he went which was the joke of the army, - nagged Belo constantly.

The dual never actually took place but we certainly revised our opinion of the Englishman before the war was over. His name was Cousens. We found that if there was any bit of dangerous work to be done, the Englishman always volunteered for it and instead of "go to it, boys" he'd say, "Come on, Follow me." It was during the Seven Days battle around Richmond, the Yanks had undertaken to burn a bridge. When the call came for volunteers to run them off, the Englishman hopped off his horse and asked us to come on and follow him. He went down that road with a six shooter in his hand, shooting across the bridge. He emptied the weapon and just stood there and reloaded the pistol by hand, taking some time to perform the operation by the methods used in those days."Cousen's went through the war without getting wounded a single time. He later married a girl from Ashland, about 14 or 15 miles out of Richmond, and I last heard from him about 10 years ago, he we still living at Ashland.

From Third Cavalry.

"I was wounded in the right leg between the ankle and knee at Chickamauga a piece of shell cut all the muscle in the leg out. I got so I could get about in three months' time. I joined my regiment about the time Grant took command of the Union Army in the East. They put me in the hospital as a nurse. I joined the army to fight, so I ran away and rejoined the command. The first time they moved they sent me back to the hospital. I left there again and returned to the outfit. Then they put me in the guardhouse and gave me corn meal and water for five weeks. One morning the doctor came and says, "How long are you going to stay here?" and I said, "Doctor has the corn meal given out?" Finally they took me out, and to keep my busy, they placed me in charge of a ward in the hospital where I remained for the duration of the war."

"Before I close this narrative," said Mr. Leach, who was one of the active members of the Hood's Brigade reunion in Bryan last week-end, and the leading story teller, "I want to state that it was the Third

Private William F. Schadt, 1st Texas Infantry.

(Rick Calvert)

United States Cavalry, stationed at the Rio Grande that furnished all the good leaders on both sides in the Civil War."

He was born in Germany and came to the United States as an infant. He was a member of the "Lone Star Rifles," later Company L of the 1st Texas Infantry. Private Schadt was captured at the Battle of Darby Town Road in 1864.

Private William Schadt
1st Texas Infantry

Dear Sister,

I take this present and only opportunity I have had to write you and inform you of the death of dear brother Charley, who fell gallantly fighting for our glorious cause on the 7th of this month. Dear Sister, I have been so oppressed by grief at the loss of our only and dear brother that I could not nerve myself to write you the intelligence of his death any sooner, but thank God has nerved me with the spirit of revenge and woe be to the hated Yankee foe that ever I meet. Dear Sister, I pray to God to nerve you in this hour of our affliction and trouble as he is the only one to comfort us in our hardest treats. Dear

Sister, do try and pray to God that he may give you courage to stand the blow that has fallen so suddenly upon you, as God is the only one we can put our trust upon you, as God is the only one we can receive. Dear Sister, Charley was decently and respectfully buried. He was laid under a large apple tree and his name and age and also the cause he fell under inscribed upon the tree. Dear Sister we attacked the enemy and drove them in confusion before us. We killed and wounded 300 and took prisoners. Our loss was 15 killed and 3 wounded. The fight Dear Sister, I received your welcome letter by Lieutenant Bedell. Goodbye, dear Sister, I am well and will write soon again. Trust in God and you will be all right.

Your affectionate brother,
William[17]

Private Alfred J. Wilson
1st Texas Infantry

Was in the battles of Eltham's Landing, Seven Days around Richmond, Second Manassas, Gaines' Mill, Cold Harbor, Sharpsburg, Wilderness, Spotsylvania, Chancellorsville, Mechanicsville, Deep Bottom and Fort Harrison, where we fought negroes twice the same day; the siege around Petersburg lasted for months and then the evacuation and retreat to Appomattox where we surrendered, stacked arms and hit the road for our own homes but was captured on the way by a fair damsel in Georgia, and was brought home to Texas and still remain in captivity.

Our first experience in battle was when our army evacuated Yorktown. It fell to the Texas Brigade to cover the retreat and after marching several days in the rain and mud, with water almost to our waists, and when near Eltham's Landing, we were met by a heavy force which had been sent up York River to head us off and to intercept our artillery and wagon train. Hood's Brigade was placed in line of battle at a point where some of Washington's old breastworks still remained and made us quite a snug hiding place. When we gotten safely in position, Col. Rainey of Palestine, Texas, paraded up and down the line, persuading the boys to keep quiet and not to fire till they could see the whites of their eyes; then aim low, and you will cripple what you don't kill. When the blue line came in sight and as they drew nearer we became so amazed at the beauty and grandeur of the scene

17. Schadt, *Letters Home*, 20.

that some of us would have forgotten what we were there for, but the ringing command of our officers which brought us to realize that this splendid array meant death and destruction and I frankly admit that some of us would rather have been somewhere else, but rallying to the situation we took aim and fired with such telling effect that it became their turn to fall into consternation and those who were not killed threw down their guns and ran for dear life, we following, shooting and giving vent to wild Texas yells. We pursued them to their gunboats and when night came on we quietly fell back out of their range. The Texas Invincibles were there with the same old guns, but after the battle we were supplied with guns, cartridge boxes and ammunition which the Federals had thrown away and our old ones were sent to the arsenal where they may be yet. The shotguns did the best execution as they were loaded with a charge called "Buck and Ball," which consisted of one ball and a number of buckshot and was wrapped in a paper and tied up so that they were convenient to carry and load and when firing always found something. It was like a charge of grape and canister on a small scale. This little affair was only child's play compared to what we were to see later on. We were nearly in all the hard-fought battles of the Army of Northern Virginia, to say nothing of the numerous skirmishes, the long hard marches with nothing to eat and a few clothes to protect us from the weather, all of which it would take a book to tell.[18]

Galveston Daily News
April 17, 1874

As originally constituted, the brigade numbered about 4,000 men. They were with Johnston in the Peninsula Campaign, and participated in the Williamsburg battle. Upon the retreat from Yorktown they brought up the rear, and skirmished daily with the enemy. Upon this retreat McClellan attempted to cut Johnston's army in two by a rapid movement from City Point; but his forces under Franklin were fought and driven back at the battle of Eltham's Landing by the splendid fighting of the Texas Brigade. In acknowledgement of services rendered at that battle, by general orders from headquarters, the Texas Brigade was allowed to carry into action, ever after, the Lone Star flag, as well as the Confederate battle flag.

18. Yeary, *boys in gray*, 803-804. Private Alfred J. Wilson was captured at the battle of Antietam on September 17, 1862. He was exchanged after the battle of Antietam and surrendered with the Army of Northern Virginia at Appomattox on April 12, 1865.

Senator Joseph D. Sayers
(Library of Congress)

Senator Joseph D. Sayers joined the 5th Texas Mounted Cavalry. Sayers was promoted to major in 1864 and was assigned to General Thomas Green's staff. He was paroled at Meridian, Mississippi on May 10, 1865. He was admitted to the Texas bar in 1866 and started to practice law. He became a state senator representing Bastrop District, Texas and was chairman of the Democratic state executive committee from 1875 to 1878. He served one term as lieutenant governor of Texas from 1879 to 1881. He was elected to the U. S. House of Representatives in 1884 and served on the committee on naval affairs. He was elected as governor of Texas in 1898 and again in 1900. At the end of his term of governor he began to practice law once again, in San Antonio. In 1911 he became regent of the University of Texas in Austin, and was chairman of the state industrial Accident Board. Sayers also served as a member of the board of legal examiners from 1922 to 1926.

Senator Joseph D. Sayers

WEEKLY STATESMAN
JUNE 30, 1887

Their Gallant Spirit.
(Excerpt of speech given to Hood's Texas Brigade Association, June 29, 1887)

Beginning at Eltham's Landing, they then and there entered upon that career of glory which they so gallantly followed through every peril and through every privation, until the coming of that hour when he, whom they loved so well, bade them fight no more.

But, let us take a brief and hurried glance at some of the memorable scenes through which you were called upon to pass.

Says Major-general Gustavus W. Smith, who commanded at Eltham's Landing, in his report of that affair: "All the troops engaged showed the finest spirit, were under perfect control and behaved admirably. The brunt of the contest was borne by the Texans, and to them is due the largest share of the day at Eltham." And we have it from a Federal general, that, had it not been for the enemy's gunboats, this would have been another Bull Run affair.

Here the First Texas lost more than two-thirds of all who fell upon the Confederate side, among the killed being Lieutenant-colonel Black

Though greatly superior in numbers, the enemy was driven back to seek the protection of his gunboats, and the retirement of the Confederates from Yorktown out of the peninsula was safely accomplished.

Their route led them along and close by a deep, navigable river filled with vessels of war, gunboats and transports of the enemy, and connected with the line of march, at almost every mile, by good lateral roads leading to favorable landings."

But this engagement, though important and successful, was but as a skirmish when compared with the heavy and sanguinary battles in which the brigade was to bear so prominent and so honorable a part.

Richmond was became an almost a beleaguered city, and the Federal commander was concentrating his troops preparatory to a general in battle, which, he confidently predicted, would result in the abandonment of the capital. A distance of only four miles intervened between the left wing of his army and the goal of his ambition. But that distance he was destined never to pass. It is unnecessary for me to enter into details as to the engagement of May 31 to June 1, 1862, known as the battle of Seven Pines.

The brigade was placed in the left wing of the army. General Longstreet, who commanded upon the right, being in need of more troops - using the language of Major General Smith, who was in charge of the left wing - the brigade moved, under orders and in double quick time, through the mud and water, underbrush, and other difficulties of the ground, to his assistance, driving in the advance pickets of the enemy upon their support, and taking and passing their camp with scarcely a perceptible halt or notice, only anxious to find the enemy, in force, who were still making resistance in front of Longstreet and Hill."

But, dark coming on," says General Smith, "there is no reason to doubt that Hood's Brigade of Texans upon the right and Griffith's of Mississippians on the left, supported by the brigade of General Semmes, would have enabled us, in one more short hour of daylight, to drive the enemy into the swamps of the Chickahominy. As it was, darkness compelled us to relinquish an unfinished task, and the troops were withdrawn from the wooded swamp immediately in contact with the enemy and bivouacked in the open field within musket range of their strong, defensive position."

As at Eltham's Landing, so at the Seven Pines do we find the brigade in the front and eager to establish a reputation for those high soldierly qualities, which they were so often and so abundantly demonstrate upon many a hard fought field before the final fall of the Confederacy. A most difficult and a most dangerous task it was to accomplish, but the character of the task was well understood and fully appreciated, and every man, with scarcely an exception, prepared himself to meet the issue.

Eltham's Landing

By Mrs. A. V. Winkler

At Yorktown, they (Texas Brigade) was assigned to the reserve corps and camped on the ground occupied by Washington's army during the Revolutionary War. A line of fortifications had been thrown up by General Magruder, and they were daily detailed to act as sharp-shooters, the Federal pickets advancing within two hundred yards of their works. Not much damage was done, as only a few were ever wounded, but they watched one another's movements with sleepless vigilance.

The evacuation of the Peninsula became imperatively necessary, from the fact that the troops were confronted by a superior force, and flanked right and left by navigable streams, occupied solely by the enemy's fleet. The Texas Brigade acted again as rear-guard from Yorktown reaching Williamsburg.

The next morning a fierce onset was made. The Federals were repulsed with heavy loss, amounting to about five thousand, killed, wounded, and missing, the Confederates about twenty-five hundred.

It became evident the next day that the Federals were only trying to retard the progress of the evacuation, and were landing troops by gunboats and transports up York River at Eltham's Landing, opposite the village of West Point, the terminus of the York River Railroad, which runs from that place, about forty miles, to Richmond. Here, too, the Pamunkey and the Mattapony Rivers unite, forming York River, and the design was to cut the Confederate army in twain right here and intercept them, while McClellan advanced upon Richmond.

General Franklin landed two regiments from his gunboats on York River at Eltham's Landing, near the village of Barhamsville, New Kent County, May 7. The Texas Brigade was marching as rear guard, and encountered the Federal picket-line which had been thrown out. General Hood immediately ordered his men to move up, which they did at double-quick, and the line of battle was formed on the brow of the hill. Beyond this hill, which had a precipitous descent, was an open field six or seven hundred yards in width. Beyond this were five or six companies of the enemy who fell back into the timber, and commenced a running fight. Another and another company was ordered to the support of the skirmishers, until six were not engaged. The Federals made a stand behind an old mill-barn, and a spirited engagement ensued. The firing became general, and the enemy many of their guns missing fire, threw them down and fled.

While the Fourth Regiment was thus engaged, Colonel A. T. Rainey, of the First Regiment, ordered his men to attack the left wing. Getting his regiment into position, they received the fire wall on an open road, the Federals in the brush. The slaughter was so great that Colonel Rainey ordered his men to fall back into the woods about one hundred yards, where they were halted and commenced to kneel and await the approach of the enemy's force. He was exceedingly nervous about whether his men or himself would stand fire, as they were all raw troops, but was compelled to appear cool and collected to inspire confidence in his men. When the regiment fell back, the Federals, supposing they were retreating, came on with a yell to within thirty steps. The Texans unflinchingly received the fire, pouring volley after volley into their ranks. After fighting about half an hour, and discovering the Federals did not advance, the Texans were ordered to rise and charge. They gave a yell and sprang forward. The Federals, seeing the situation, turned and fled to their gunboats, about four or five hundred yards distant, with the Texans in full pursuit.

General Hood arrived with his staff about then. Perceiving that they were about to run under the fire of the gunboats, a courier was dispatched to order Colonel Rainey to halt. They did not obey the order. General Hood himself now came up and ordered, "Colonel Rainey, halt your regiment!" This order was obeyed. In this action the Texans engaged were about seven hundred. Federals, eighteen hundred or two thousand. Lieutenant Colonel Black, Captain Decatur, and twenty privates were killed and some thirty or forty wounded. The Federal loss was three hundred killed and wounded, and one hundred and twenty six prisoners, according to General Hood's official report.

This affair was of great importance. President Davis in conversation with a Texas Senator, said, "They saved the rear of the army and the whole of our baggage-train." General Gustavus Smith, in a letter to Horace Randall said, "The Texans won immortal honor for themselves, their state, and their commander, General Hood, at the battle of Eltham's Landing at West Point."[19]

Anderson County In The Civil War

On April 29, 1861, Jefferson Davis spoke before a special meeting of the Confederate Congress in response to Lincoln's call for troops two weeks earlier. This excerpt is from his speech: "The Declaration of War made against this Confederacy by Abraham Lincoln, President of the United States, in his proclamation issued on the 15th day of the present month, renders it necessary, in my judgement, that you should convene at the earliest practicable moment to devise the measures necessary for the defense of the country."

Every southern state was hastily making preparations for war. Even distant Texas, which had seceded on May 2, 1861, had immediately acted and had secured the surrender of all Union troops in Texas and their supplies without a single shot being fired.

19. Winkler, A. V., "Hood's Texas Brigade" in Dudley G. Wooten, ed., *A Comprehensive History of Texas 1685-1897*, Vol. II, Dallas, TX: Book, 654-655.

Jefferson Davis, President of The Confederate States of America.

(Library of Congress)

In Anderson County, the fever of war was high. The county had voted 1,500 to 7 in favor of secession and had been represented in Austin at the state Secession Convention by the capable John H. Reagan, A. T. Rainey, and S. G. Stewart. T. J Word represented us in the adjourned convention.

John H. Reagan was chosen by the Texas convention to attend the general convention in Montgomery, Alabama. Upon arriving, Reagan was called upon by Jefferson Davis and was persuaded to accept the cabinet post of Postmaster General. Reagan kept his office until the last few days of the war when Davis, clinging to a lost dream of saving the Confederacy, appointed him to the post of Secretary of the Treasury.

Shortly after this, Union troops captured Davis and his escort, including Reagan. Years after the war, it would be John H. Reagan's account of the capture that disproved the rumor that Davis had tried to disguise himself as a woman to avoid detection.

Between 700 and 1,000 men from Anderson County alone would march off to war, many to never return.

Dr. John Woodard of Palestine organized the first full company that was completely made up of men from Anderson County. This was Company G, First Texas Regiment. When this company left Palestine,

John H. Reagan, Postmaster General of The Confederate States of America.

(Library of Congress)

"their uniforms at first did not consist of the Confederate grey, which was substituted later, but of a dark suit with red stripes. It was at Manassas that this company first saw action, and it was there that Dr. Woodard was killed.

In the summer of 1861, Colonel Lewis Wigfall began organizing companies in East Texas. A. T. Rainey began organizing a company here in Palestine. Company H was assigned to the First Texas Regiment along with Dr. Woodard's Company. The First Texas Regiment was joined by the Fourth, and Fifth Texas Regiments plus a Georgia Regiment. Together they were called the First Texas Brigade, under the command of Wigfall; however, after Wigfall was voted to the Confederate Congress, he was replaced by John B. Hood. Thus began a legend that stands beside that of the Light Brigade; Hood's Texas Brigade became world renown and people everywhere shared the same regard for these Texans that General Robert E. Lee felt when he stated to an observer while the brigade marched in review: "Never mind their raggedness, the enemy never sees the backs of my Texans!" Men from Anderson County helped write that legend.

At Eltham's Landing, Rainey's company felt their first taste of gun fire. The First Texas Regiment of which Rainey was now in command was being held in reserve by General Hood, when they were suddenly fired upon by Union troops. Rainey ordered his men to move back into the trees and "...fall on your knees, unfurl the Lone Star Flag, aim low, and give them hell." The Union advance was stopped some 30 yards from Rainey, and after 20 or 30 minutes of close fighting, Rainey ordered a charge and with a wild Rebel yell, the First Texas leaped to their feet and drove the enemy back. They did not cease until General Hood himself, who had been leading the rest of the Brigade elsewhere on the battlefield, rode out and ordered the First to half and retire, leaving the Federal troops in complete confusion.

At the Battle of Sharpsburg (Antietam), Hood's Brigade was almost completely destroyed on the battlefield. Ordered to hold a position with little ammunition against one entire wing of the Union army, Hood's Texans did just that. Heavily outnumbered and battle weary, these men held their ground and repulsed attack after attack until the battle died down. When asked by General Lee on the condition of the Texas Brigade, Hood replied "The Texas Brigade is dead on the field." When the casualty reports were compiled at the end of the war, it was discovered that at Sharpsburg Hood's Texas Brigade had lost 82% of her fighting force in the battle, and that the First Texas Regiment of which Rainey and Woodward's companies had been assigned had lost close to 85%, the highest casualty rate inflicted on any fighting outfit during the four years of the war.[20]

20. *Anderson County In The Civil War*, in *The Tracings*, Palestine, TX: Anderson Historical Society, 1986, Vol. 5, Number 1, Winter 1986, 4-7.

4th Texas Infantry Flag flown at Eltham's Landing.

(United Daughters of the Confederacy - Texas Division)

CHAPTER 2

4th Texas Infantry Regiment

Company A: (Hardeman Rifles) Goliad County – J. C. G. Key
Company B: (Tom Green Rifles) Travis County – B. F. Carter
Company C: (Robertson Five Shooters) Robertson County – W. P. Townsend
Company D: (Guadalupe Rangers) Guadalupe County – J. P. Bane
Company E: (Lone Star Guards) McLennan County – E. D. Ryan
Company F: (Mustang Greys) Bexar County – E. H. Cunningham
Company G: (Grimes County Greys) – J. W. Hutcheson
Company H: (Porter Guards, Walker County – P. P. Porter
Company I: (Navarro Rifles) Navarro County) – C. M. Winkler
Company K: (Sandy Point Mounted Rifles) Henderson County - W. M. Martin

Colonel John F. Marshall, Commanding officer of the 4th Texas Infantry

(Texas Heritage Museum Historical Research Center)

The citizen soldiers of the 4th and 5th Texas Volunteer Infantry Regiments were primarily recruited in Central Texas under a levy imposed on the state by President Jefferson Davis on 30 June 1861 that called for 2,000 troops. An Austin newspaper editor, John Marshall, later command officer of the 4th Texas Infantry, was convinced that most of the fighting would take place in the East and he made his plea with Davis so that Texas would have a sizable representation.

Major William P. Townsend, 4th Texas Infantry.

(Texas Heritage Museum Historical Research Center)

Colonel John Marshall, Commanding officer of the 4th Texas Infantry. Before the war he was editor of The Austin State Gazette. When the war began, he left journalism and joined the Confederate army as a private. He was quickly promoted to Lieutenant-Colonel and became the commanding officer of the regiment, when John Bell Hood, former commanding officer of the 4th Texas was promoted to brigadier general. Colonel Marshall was killed leading the 4th Texas on their charge towards the Federal lines, during the Battle of Gaines' Mill on June 27, 1862. He is buried at the Confederate cemetery in Richmond, Virginia.

Major William P. Townsend[1]
4th Texas Infantry

Chickahominy River 15 Miles
from Richmond, May 11, 1862

Dear -,

We are hourly expecting the big battle of the war. The enemy are in two miles of us – our regt. is nearest him.

On the 7th we had a considerable brush with him. He had landed heavy forces from the York River to cut off our Brigade which was in

1. Major William P. Townsend became the original captain of the Robertson Five Shooters, which later became Company C of the 4th Texas Infantry. From February to April of 1862, he was detailed to recruiting duty back in Texas. He was promoted to Major on July 10, 1862. He was wounded in the foot at the Battle of Second Bull Run (Second Manassas) on August 30, 1862 and later had his foot amputated. He was appointed on rotary court-martial on General William J. Hardee's Corps during the winter of 1862-63.

Captain John "J. W." Hutcheson, 4th Texas Infantry

(American Civil War Museum under the management of Virginia Museum of History and Culture)

the rear - We drove the Yankees back to their gunboats like so many whipped hounds. It was purely a Texas fight- no other Confederates engaged. Our loss was 11 killed and 25 wounded. None from my company.

General Johnston has complimented the Brigade highly and the Texas stand No. 1 in the estimation of all. You never saw such cowards the Yankees are.

I think the battle cannot come off under three or four days. We will probably have skirmishing in the meantime. Since I have reached the army I have had but one chance of writing to you. We have sent off all our baggage. I have nothing but the clothes I have on and one small blanket - never take off my boots at night and pack - in our haversacks- But all are still in the highest spirits and are as confident of success as you ever way anyone...We have good news from Stonewall Jackson, he has soundly thrashed the Yankees and Beauregard seems to be doing the same with (?) When we feather McClellan's Army - you may look for your Notary and a whisper in my ear tell me that it will not be long...

Dearest kick up your spirits for our cause and is on the hands of the Lord..."[2]

Captain J. W. Hutcheson was a member of the Texas Secession Convention of 1861 and was a signer of the Texas Secession Ordinance. When the war broke out he joined the "Grimes County Grays," which became Company G of the 4th Texas Infantry Regiment. He was quickly promoted to captain. Captain Hutcheson was mortally wounded at the battle of Gaines' Mill on June 27, 1862, and died two days later on June 29, 1862.

2. William Townsend to unknown, May 11, 1862, Hood's Texas Brigade Files, Texas Heritage Museum-Historical Research Center.

Captain J. W. Hutcheson
4th Texas Infantry

HOUSTON TELEGRAPH
JUNE 2, 1862

Army of the Peninsula
Bivouac, 4th Texas Regiment
On the Road Near New Kent Courthouse, Va.
May 14, 1862.

Editor Telegraph –

On Wednesday the 7th instant, near Barnhanville, on the road leading from Williamsburg to New Kent Courthouse, ten miles from the latter place, the Texas Brigade for the first time had a brush with the enemy, A sketch of the entire affair would no doubt be interesting to your readers, but I can only give you an account of the part of our company, "Grimes County Greys," took in the engagement, leaving it to other companies, that acted in different parts of the field, to speak for themselves. No better description of the affair can be given than the report of Captain Hutchison, and having obtained a copy of it I take the liberty of sending it to you:

Headquarters 4th Texas Regiment
May 19th, 1862.
Col. Marshall, Commanding 4th Texas –

Sir: In compliance with our order I herewith furnish a report of the part acted by my company in the engagement of the Texas Brigade with the enemy near Barnhanville, the 7th inst.

Soon after the Regiment had been drawn up in line of battle and the company of Capt. Carter had been thrown out as skirmishers in front, my own was ordered to support him and if necessary, to deploy to his right. After advancing through the woods at supporting distance from him for about 100 yards, by order he deployed to the left, and I to the right of the road near which the engagement took place.

I soon found a marsh covering a great extent of the ground on which my deployment was to be made; and the tracks of individuals, supposed to be the enemy, retiring from the front of the regiment advised me to proceed with the utmost precaution, and more especially, as ahead of us protected too by the trees of the forest, the enemy might securely hide,. Creeping upon the hands and knees we reached this dam, where there were various indications to a woodman, that the enemy was near, here, and where we could see over the dam, I ordered a halt, and for some time carefully surveyed the ground before us. By careful and close scrutiny we discovered the enemy behind the trees, form which he occasionally fired. Sliding to position where he would be unmasked, my company began to open a fire, which, proved hot and affective, the enemy soon became quite visible retreating and skulking away. Forward was ordered, and closely pursued, here and there, one, and sometimes more, until at one place seven were found weltering in blood, being struck down by the cool and unfailing aim of the men. Thus it continued for some time, in an almost constant fire along the edge of this lake, until the whole hill was strewn with the dead and wounded of the adversary – we pressing him and never giving him time to rally or hide effectually.

We followed them up until we met with the 5th Texas skirmishers, then were ordered to return to the regiment then forming, below us, where the 1st Texas had just had an encounter with the enemy.

This closed my separate connection with the fight. Beyond doubt we took in the skirmish, 21 prisoners, and the killed and wounded could not have been less than 35 and most of these belong to the 1st class. Among the killed was a Captain, Lieutenant and sergeant. We took a number of small arms, among them three officers swords.7 Nothing could surpass the coolness of my men under this, their first fire, and the fatality and certainty of their aim was incredible. We lost none killed and but one, C. W. Spencer, seriously wounded.

I will mention that Maj. Warwick was with us and brought up my reserve (20 of the company being left behind for that purpose) handsomely and well, I will also call attention to the fact that several parties being sent out on this ground to skirmish independently, and the woods and localities being unknown to us, without any well-defined mode of recognition, exposed to an encounter with friends, which required great precaution, exposed to an encounter with friends, which required great precaution to prevent, and which rendered the undersigned painfully uneasy.

I have the honor to be your obd't serv't.
J. W. Hutcheson, Capt. Company G. 4th Texas Regiment.

GALVESTON DAILY NEWS
MAY 8, 1874

The first year of the war was not of much genuine active service for the Texas brigade. The armies of both the North and South, after the affairs at Bethel and Manassas, feeling the magnitude of the struggle upon which they had engaged, lay watching each other like gladiators. The celebrated pronnciamento of Mr. Lincoln, ordering the dispersion of Confederate troops, failing in its effect, and the recoil of the first army sent out from the North upon the National capital, taught that section that the ninety-day business was a mistaken calculation; but true to the genius of the people of that portion of the Union, they went to work at permanent organization of their military departments. McClellan was then chief in command of the United States. Few will deny the capacity of this officer as an organizer. He masterly genius taught him that a people endowed with the courage and high chivalric qualities of the Southerner will require more demonstrative arguments than paper bulletins to desist from their undertaking; and as a consequence, he employed the fall and winter of 1861 and the spring of 1862 in the discipline and armament of several splendid army corps. Magnificently equipped with the best arms then known to the service, with artillery in abundance, and a commissariat affording every comfort and necessary to the troops in the field. McClellan made a sudden change of front from the line of the Rapidan and turned up at Fortress Monroe. The chief with which the Federal was then contending was the able and sagacious Gen. Joseph E. Johnston. This movement of McClellan forced Johnston to counter action, and the scene of warfare was transferred for a time to the Virginia peninsula.

The rapid "falling back" of Johnston's army upon the line of invasion, was one of that great chieftain's most splendid achievements. It blinded and confused the Federal leader. When McClellan touched the Southern lines at Yorktown, stretching to the James river, he found a handful of men under the hazardous Magruder, and never dreamed that such a force would contest the road to Richmond with an enemy of one hundred and twenty thousand strong. He fancied that Johnston's forces were on the ground, and set himself to work at siege operations. The result is a matter of history. The Confederate army of the Rapidan had time to reach the peninsula before McClellan had time to reach the peninsula before McClellan was aware of his mistake, and Richmond was saved for further and more desperate contests.

The arms that for a year before had remained unused had no further time to spare. Among the first troops to reach the battle grounds were those comprising the Texas brigade. The three Texas regiments were armed with the minie rifle - we think the Eighteenth Georgia had only the smooth-bore musket. At West Point, or, Eltham's Landing, the brigade first stretched its maiden sword. And a fine piece of steel it was. No discount on the boys from the prairies that morning.

3rd Lieutenant William E. Barry, 4th Texas Infantry.

(Confederate Veteran Magazine)

3rd Lieutenant William E. Barry
4th Texas Infantry

Three Glorious Regiments
Eltham's Landing

Our first regular battle occurred on the 7th day of May, 1862, as what is known in history as Eltham's Landing on York River, and by some as West Point. The army, under command of Gen. Jos E. Johnston was on the march from Yorktown to Richmond, followed by Gen. McClellan and the Yankee army. Early in the morning the brigade was quietly marching on the road, bringing up the rear of the Confederate army. When on top of a high hill we were fired upon very suddenly from and apple orchard at a distance of not over 50 yards, the Fourth Texas being in advance of the brigade with General Hood at the head of the column. We were marching according to orders with unloaded guns - only one man, John Deel of West Texas, had his gun loaded; he quietly fired and brought one of the Yankees to the ground out of an apple tree. Only one man was wounded by this fire in our ranks - Hart T. Sapp, now residing in Houston - who has the honor of being the first man belonging to the brigade wounded in battle. Although the fire upon us was sudden and unexpected, there was no panic among the Texans, who proceeded at once to load their rifles and fired a volley into the squad of Yankees, killing four of them before they could escape to the timber.

Company G from Grimes county, and Company B from Travis County, two very large companies, was thrown out as skirmishers and advanced down 600 yards through an old field towards the dense woods in the bottom lands of York river. We fully expected to receive a volley from the enemy just before we entered the timber, but not a gun was fired at us.

We discovered just within the timber the four dead Yankees. As we advanced we became actively engaged fighting from tree to tree. Steadily advancing, the three Texas regiments became heavily engaged, and more particularly the First Texas, which had a fearful struggle with the First California. After repeated charges and repulses they succeeded in driving the First California under the protection of the gunboats in the York river. The engagement between Franklin's Division and the Texas Brigade. History devotes but scarcely half a page to this battle. Its successful result saved the army wagon trains with supplies and munitions of war belonging to the Confederate army, for which we received the following thanks and praise from Gen. Gustave A. Smith, commanding the division:

"The Texans won immortal honor for themselves, their State, and for their commander, General Hood, at the battle of Eltham's Landing near West Point. With 40,000 such men could not hesitate to invade the North, and would before winter makes them sue for peace upon our terms, or destroy their whole country. But in praise of the Texas Brigade of my division, I could talk a week, and then not say half of what they deserve. If the regiments now organized in Texas, could be transported here and armed tomorrow, properly led, they would end the war in three months.[3]

1st Lieutenant William T. Hunter was first enlisted as a private. When Texas seceded, he helped recruit for the "Porter Guards," later designated as Company H, 4th Texas Infantry. His company elected him as a 3rd Lieutenant.

3. Chilton, Unveiling, 102-103. The excerpt about the Battle of Eltham's Landing is part of a speech given by Lieutenant Barry during the Hood's Brigade Association annual reunion in Navasota, Texas on June 27-28, 1907. 3rd Lieutenant William E. Barry originally enlisted as private in the "Grimes County Grays," which later became Company "G," 4th Texas Infantry. He was wounded in the arm at the battle of Gaines' Mill on June 27, 1862, and was captured at the battle of Sharpsburg on September 17, 1862. He was confined at Fort Fisher POW camp and exchanged on November 10, 1862. Soon after being exchanged, he was elected by his company as 3rd Lieutenant. He was wounded again at the battle of the Wilderness on May 6, 1864.

1st Lieutenant James T. Hunter, 4th Texas Infantry

(Confederate Veteran Magazine)

1st Lieutenant J. T. Hunter
4th Texas Infantry

At Yorktown In 1862 And What Followed.

When we got to Williamsburg, just after passing through the fortifications, a staff officer came in great haste to General Hood with an order. General Hood turned in his saddle and said to us: "Quick time! March!" We marched all day until after dark, were then halted, and slept in fence corners in a lane. At break of day the order to "Fall in" was given, and we hurried on until about ten o'clock, when we were marched into a thick timber and told to lie down and rest, but to keep very quiet.

To explain as best I can the object of this move: At this point the James and Pamunkey Rivers are only seven miles apart, and General Johnston supposed that McClellan would try to cut off his retreat by hurrying transports up the Pamunkey River to Eltham's Landing, opposite to where we were then, and establish a line across between the river's; and the event which followed proved the sagacity of General Johnston's judgment. Of course General Hood took every precaution; had scouts on the river and pickets between us and the river. The transports with General Franklin's corps arrived that night and disembarked. The next morning Hood's Brigade was formed and marched toward the river on two different roads, the 4th Texas and the 1st Texas on one road and the 5th and Hampton's South Carolina Legion on the other. The- 18th Georgia followed the 1st Texas and was held in reserve to guard Captain Reily's battery of artillery. After the 4th had proceeded about one mile, we came to a high, open piece of land from which a steep declivity overlooked the river valley. From the base of this hill for about six hundred yards was a field, and the side of the hill was grown up in bushes; on this hill and not far from its margin stood a house at which one picket was stationed. General Hood and staff had just ridden past the house to the edge of the hill to look over the valley, when a square of some twenty or twenty-five Yankees rose up from among the bushes and fired on the party. None of the party was hit; but H. T. Sapp, of my company, some five hundred yards in the rear, was struck over his right eye by a partially spent ball. I was walking just behind him reading a pamphlet someone had dropped. He fell just at my feet, and, seeing the blood spurt from his forehead, I, of course, thought him killed. I picked him up, laid him outside the road, and ordered the men to load their guns (as we were still inside the pickets, no guns were loaded). It so happened that John Duran, of Company A, which was in front, had his gun loaded, against orders, and the sergeant in command of the Yankees, instead of running, as his men did, stood reloading his gun when Duran shot him.

Right here was some of the best shooting I ever saw. The Yanks, after running across the field, believing they were out of range of our guns, stopped in bunches in the fence, corners; and when some twenty or twenty-five of Company A got their guns loaded, they fired on them and killed eight, and the distance was at least six hundred yards. We were armed with the Springfield rifle musket, a fine long-range gun.

General Hood ordered Company A forward as skirmishers and followed with the regiment. Soon after getting into the bottom, as the enemy's strength developed, he pushed forward more companies, taking them first from the right and then the left alternatively. Mine was the "color" company and occupied the center; hence it was the last to go into action. By this time a strong force had developed on our extreme left with the intent of turning our skirmish line and getting in our rear, and I was sent in to meet the force and stop this move and did meet and fight this regiment to a standstill. They reinforced and kept encircling us until my men could not protect themselves behind trees, and I was forced to give ground, but fell back in order without any haste. We had retired but a short distance when we came on the 1st Texas lying down. I formed my men on the right of the 1st, but was anxious to sec General Hood. I was a young soldier then, and this was the first fight in which I had commanded the company (Captain Porter was acting as major that day), and I did not know that General Hood would think I had stayed and fought long enough. Very soon I saw him coming, walking down in the rear of the 1st. I went to meet him, but before speaking my fears were allayed, for I saw that those twinkling, expressive eyes were laughing. I told him I had been fighting in front, but they got so far around me I could not protect my men and had to fall back. He laughed and said: "Lieutenant, I was watching you. You made a good fight and held those people until I brought the 1st Texas and laid them down behind you." I told him I had formed on the right of the 1st, and he said: "That's right; here is where the fight is going to be. You just obey orders given the 1st."

Very soon the Yankees came, seeming to think there were none to confront them. When they got near enough, the order to fire was given. A full volley poured forth, and every man sprang forward in an impetuous charge with that terrifying Rebel yell, and those Yanks never stopped running until they were aboard their boats. Then they opened with their artillery, shooting entirely over us, seeming to think we were retreating. We were lying near the bank.

This ended the fighting and General Hood had ambulances come in and take out our dead and wounded. Our casualties were comparatively very light. We had seventeen killed and twenty-four wounded, while the enemy reported eight hundred killed. They fought one full division of Franklin's Corps, while we fought only Hood's Brigade, composed of three Texas regiments, Hampton's South Carolina Legion, and the 18th Georgia Regiment, and it was not engaged. Our Fight was made in skirmish formation, except the action of the 1st Texas at the close. General Hood saved all the spoils of the victory. General Johnston said: "This battle is the most important in results of any up to this time. If they had succeeded in their attempt. I could probably have saved my army, but most likely would have lost my transportation." After taking care of the dead and wounded and all the spoils of the victory. General Hood assembled his command and quietly marched off; and after reaching the main road and waiting until the last of General Johnston's army passed, we again took our place to protect the rear. We marched all night and until afternoon of the next day, when, as McClellan abandoned the pursuit, our command was halted to give the men much-needed rest and something to eat, as we had had very little since leaving Yorktown.[4]

Chaplain Nicholas A. Davis served as chaplain of the 4th Texas Infantry. As well as giving sermons, he took care of the wounded and dying in Hood's Texas Brigade. He believed the soldiers from Texas were not receiving enough praise and recognition, thus he wrote, Campaign from Texas to Maryland, with the Battle of Fredericksburg. The book was published in Richmond in 1863, and a slightly revised edition in Houston in 1864.

Reverend Nicholas A. Davis
4th Texas Infantry

Battle of Eltham's Landing

The command was put in motion at daylight of May 7th, and about 7 o'clock A. M., came upon a picket of the enemy, who fired two shots at Gen. Hood, who was riding at the head of the 4th Texas, now in front. One shot struck Corporal Sapp, of Co. H, in the head inflicting a severe but not dangerous wound. Private John Deal, of Co. A, whose gun was loaded, immediately fired upon the pickets as they ran, and struck the only one in sight, killing him instantly. Some confusion was observed at first in consequence of empty guns, and Col. Marshall's order to "Fall back into the woods and load;" but Gen. Hood immediately called out to the men to "move up," which they did at double quick, and line of battle was immediately formed on the brow of a hill. Beyond this hill, which had a precipitous descent, was an open field of six or eight hundred yards width. On the opposite side were some four or five companies of the enemy, who immediately began falling back into the timber, but not until several random shots had been fired by our men, which we afterwards discovered had killed five and wounded as many more. Company B (Capt. Carter) was then ordered by Gen. Hood to deploy as skirmishers and "fell the enemy." They began a "running fight." Co. G. (Captain Hutcheson) was then ordered forward to support Co. B, if necessary; if not, deploy on its right – the latter course was adopted. Co. K (Capt. Martin) was next sent to support Co. B, and Co. E, (Capt. Ryon) to the support of Co.

---

4. J. T. Hunter, "At Yorktown in 1862 And What Followed" in *Confederate Veteran*, Vol. 26, March 1918, 66-67.

G. After retreating about half a mile, the Yankees made a stand behind an old mill-dam, and a spirited engagement ensued between them and the right platoon of Co. B, under Captain Carter, and Co. G., Captain Hutcheson - Co. H (Capt. Porter) now arrived upon the ground, with orders to support the left platoon of Co. B, under Lieut. Walsh. The firing new became general, and the enemy, many of their guns missing fire, threw them down and fled. While pursuing them, the second platoon of Co. B came upon a large force (some two hundred,) protected by a heavy palisade. This was more than was bargained for, and the boys, some twenty-five in number, immediately "treed," and answered their volleys, by picking off every one who showed his head. At this juncture Gen. Hood appeared, and ordered the Lieutenant in command to charge the works, and he would send support. Just as the command "charge" was given, and the boys with a yell, had just started for the works, the first platoon of Co. B appeared upon the left flank of the palisade, and the Yankees fled in confusion, leaving seventeen killed and several wounded in the track of their flight. While Co. B was thus engaged, Co. G had also had its share of "fun." Discovering a company of about eighty Yankees, Capt. Hutcheson with his company and part of Co. E, attacked them so vigorously, that they dared not run and were so unnerved, that they fired volley after volley into the tree-tops. Capt. Hutcheson, who was a Chesterfield in manner, did not for a moment forget himself during the fight. "Charge them, gentlemen, charge them." "Aim low, gentlemen, aim at their waistbands," were his constant exhortations, until a portion of the enemy cried for quarters. "Throw down your arms gentlemen, you scoundrels, throw them down." Sixteen obeyed the order, and the remainder taking advantage of the momentary cessation of hostilities, turned and fled. Bewildered, however, they took the wrong direction, and coming upon the 5th Texas where it was

Chaplain Nicholas A. Davis, 4th Texas Infantry.

(Texas Heritage Museum Historical Research Center)

lying down in line-of-battle, they were greeted by a volley, which left not one standing. The fruits of Capt. H's victory, were eleven killed, several wounded, and sixteen prisoners, together with several stand of arms. While these events were transpiring, the 1st, 5th, and remainder of the 4th Texas had entered the timber, leaving the 18th Georgia to support the artillery in the rear. A Yankee regiment now appeared upon the left and rear of the skirmishers, with the intention, doubtless, of cutting them off. Here we witnessed for the first time,

The Gallantry of the First Texas

The regiment now advancing, 1st California, evidently intended to fight well, and advanced steadily to within eighty paces of the 1st Texas, when they halted, poured a volley, and with three huzzahs, attempted to charge. This was expected, and "aim low, fire," was ordered by Colonel Rainey, and a discharge followed that seemed to mow down the whole front flank, and sent the remainder in confusion back again. A whole-souled hearty yell now went up from the Texans, such as only Southerners can give, and they in turn, charged. But the Californians were not yet ready to yield, and rallying, they made a stubborn resistance, and for about twenty minutes, the fire raged with terrible fury. The Texans charged again, and the enemy broke and fled, leaving about two hundred killed and wounded on the field, and several prisoners in our hands. The loss of the 1st Texas in this engagement was eleven killed and twenty-one wounded. Among the former, however, we regret to chronicle Colonel Black and Captain Decatur, who were loved and mourned by all, as brave men.

After the rout of this regiment, the enemy did not again attack us, but contented themselves with shelling us from their gunboats, and sweeping the woods with grape, from a battery they had planted upon the river bank, without, however, doing us the slightest injury. While this was going on, the boys had a hearty laugh at the conduct of an Indian Warrior,

Who was attached to the 1st Texas Regiment. During the entire battle, with musketry, he conducted himself in the most gallant manner, and had even succeeded in capturing a Yankee, whom he turned over to the proper officer, with the brief announcement, "Major, Yank yours, gun mine," and again participated in the struggle. When the first shell came tearing through the tree-tops, with its screaming inquiry, "where you, where you," he uttered a significant "ugh!" and listened until it burst. At that instant, another came, and exploded just over our heads, when he sprang to his feet, exclaiming, "no good for Indian," and made for the rear with the agility of an antelope. The boys did not, however, reproach him, because it has long been understood that Indians won't stand to be shot at by wagons, more particularly when the projectile itself shoots so terribly. The entire loss of this engagement was thirty-seven. Of that number, Captain Denny, Commissary of the 5th, was killed by a picket, and two men captured, as previously stated. Corporal Sapp, of Co. H, and Private Spencer, of Co. G, 4th Texas, were wounded, all the other casualties were of the 1st Texas, of which regiment, we cannot speak too highly. – These are the men who came from their distant homes, at their own expense, before the President had called upon Texas for troops, to assist in this great struggle. And, though their names have not occupied a place in the journals of the day, they have ever been at their posts, ready and willing to do and die for our common cause. They are a lively, merry set, and though often hungry and "ragged," they have shown in numberless instances, that they can march as far, and fight as hard, as any troops in the service.

The Enemy's Loss

In this engagement, as estimated by General Hood in his official report, was three hundred killed and wounded, and one hundred and twenty-six prisoners. McClellan's estimate is even greater, as he reported a loss of five hundred men and officers. This is probably correct, though a New York paper, which claims that the troops participating were chiefly from that section, viz: Albany, states the entire loss at twelve hundred. A correspondent of the New York "Herald," writing from West Point soon after the fight, gravely asserts that they "were charged furiously by four regiments of negroes!" This paragraph caused considerable sport among the boys, being regarded as a direct reflection upon the state of the brigade toilet. The writer, however, was in probability, more knave than fool, for just at that period, the question of enlisting slaves in the United States army was being agitated, and such an assertion would not be without its effect on the unthinking masses of the North.

The Importance of This Battle.

In reference to which, the Richmond papers have been silent, cannot be better illustrated than by reference to the language of some of our general officers. President Davis, in conversation with one of our Senators said, in speaking of the Texas Brigade, "they saved the rear of our army, and the whole of our baggage train."

General Gustavus W. Smith, in a letter to Colonel Horace Randall, writes, "the Texans won immortal honor for themselves, their State, and for their commander, General Hood, at the battle of Eltham's Landing, near West Point. With forty thousand such men, I would not hesitate to invade the North, and would before winter, make them sue for peace upon our terms, or destroy their whole country. But in praise of the Texas Brigade of my Division, I could talk a week, and then not say half they deserve. If the regiments now organized in Texas, could be transported here, and armed tomorrow, properly led, they would end the war in three months.

General Samuel W. Melton, also writes, "here we first had a fair sample of your Texans, under Hood. They are, incomparably, the best fighters in the Confederacy; men upon whom one could depend under all circumstances – who seem to fight for the very love of it. Oh! that we had more of them. Forty thousand such men could march through Yankeedom now, from one end to the other, and conquer a peace in a month."

The Brigade "Cuts Dirt," while the Yankees Dig.

The fighting ended at 2 o'clock P. M., and the enemy showing no disposition to leave their gunboats again, the brigade was ordered back from the bottom, leaving only a sufficient force for observation. Returning to the camp, from which we had started in the morning, we remained until 10 o'clock at night, when the whole army, baggage and all having passed up the road, we again assumed our position as the rear guard. Strict silence and quick time being enjoined, I am sure no troops ever marched more swiftly, or kept more obstinate silence than we did until daylight. How ludicrous the scene. What a hearty laugh a man could have had, had he been in a position to observe both armies that night. Ours, moving swiftly and steadily along, casting many and anxious glances to the rear, fearing to discover the head of a pursuing column – theirs digging, toiling and sweating, in preparing to receive the furious onslaught which they knew the "rebels" would make at daylight. Then to have watched the Yankees in the morning, feeling cautiously through the woods, listening every moment for the dreaded sound of the guns of troops, were miles on their way to Richmond, and still going. Late in the afternoon of May 8th, the brigade was drawn up in line of battle, in the lawn, in front of Doctor Tyler's residence, five miles west of New Kent Court House, as the enemy were threatening to attack us. They did not, however, come up, and we remained here until the following evening, when we moved one mile up the road, and formed a new line of defense, to be held until our army could reach, and take its position in front of Richmond. About noon on -----, we decamped, and, though constantly in motion, only reached the Chickahominy, about six miles, by 1 o'clock at night. This was owing to the fact, that the road was blocked up by the rear of our artillery and baggage train, and not daring to lie down or rest, we could only "mark time" in the rain and mud until the hour above mentioned, when all others having passed over, we reached the bridge. Here we found several Generals, with their attendant aide and couriers, all exhorting us to "close up," and for God's sake hurry. This was more easily said, than done, for the roads had been cut by artillery and wagons, until a perfect mortar had been formed from one to three feet deep, and through this below, and a heavy soaking rain above, the men floundered on. At length, losing all patience, General Whiting dashed upon the bridge. "Hurry up, men, hurry up, don't mind a little mud." "D'ye call this a little mud! s'pose you git down and try it, stranger; I'll hold your horse." "Do you know whom you address sir? I am General Whiting." 'General ------, don't you reckon I know a General from a long-tongued courier?" says the fellow, as he disappeared in the darkness. This, repeated with sundry variations several times, at length discouraged the General, and leaving the Texans, whose spirits he had threatened to subdue, to cross as best they might, he rode away. Finally all were safely landed on this side the Chickahominy, and without waiting to eat or build fires, the men threw themselves upon the muddy ground, and slept soundly until morning. We occupied this point until evening, and then moved back about two miles, and bivouacked until the command was relieved, and marching to the rear, we camped at "Pine Island," three miles east of the city. – Nothing of interest occurred here. The men gave their whole attention to eating, sleeping, washing bodies and clothes, and watching the recruits who had recently arrived, attempting "balance and left." On Sunday, the Chaplain, having just returned from Texas, where he had gone on recruiting service, we had Divine worship, which was remarkably well attended.

Preparations For the March – Again.

May 26th. – Orders were issued to send off surplus baggage, which always communicates with amazing facility when the cam is near a city or town. On the evening at sunset we departed, and marching and "marking time" all night, we accomplished a distance of seven miles,

Sergeant Val C. Giles
4th Texas Infantry

(Dolph Briscoe Museum of American History University of Texas at Austin)

and at dawn were halted one mile this side Chickahominy, on the Meadow Bridge road. Here we remained until the following day concealed in the woods, and then marched back and camped between the Mechanicsville Turnpike and Central railroad. On the next evening a most terrific thunderstorm, accompanied by torrents of rain, began and lasted through the night, thoroughly drenching the men. One man in the 4th Alabama Regiment, camped near us, was killed by lightning, and several were severely shocked.[5]

Sergeant Val C. Giles kept a journal of his war experiences in a book titled *Rambling Reminiscences of the Stormy Sixties*, which became a popular read after its publication. In 1961, Mary Lasswell edited his journal which she titled, Rags and Hope, The Recollections of Val C. Giles, Four Years with Hood's Brigade, Fourth Texas Infantry, 1861- 1865.

Sergeant Val C. Giles

4th Texas Infantry

GALVESTON DAILY NEWS
SEPTEMBER 21, 1902

From The Potomac To the Peninsula.

When General Joseph E. Johnston evacuated Manassas in the early spring of 1862 the Texas brigade was still in winter quarters at Dumfreze on the Potomac.

5. Davis, Nicholas, *The Campaign From Texas To Richmond, With The Battle Of Fredericksburg*, Richmond, VA: Office of the Presbyterian Committee Of Publication of the Confederate States, 1863, 32-35.

We had spent the winter there drilling, building log cabins, picketing the Potomac River and hunting 'possums on the yellow persimmon hills.

On the evening of March 7, 1862, we received the order (afterward so familiar to us all,) "Cook three days' rations and be ready to move at a moment's warning."

The winter had been extremely severe, the hills and valleys having been covered with snow for nearly six weeks. About March 1 the weather began to get warmer and the thaw set in. The earth was soaked and all the water courses were brim full and overflowing.

On the afternoon of the 8th the regiment was formed and Colonel Hood made us a speech. Riding to the center of the regiment, he found the men, took off his hat, looked up and down the long line, then said:

"Soldiers! I had hope that when we left our winter quarters it would be to move forward, but those who had better opportunities of judging than we have ordered otherwise. You must not regard it as a disgrace, it is never a disgrace to retreat when the welfare of your country requires such a movement. Ours is the last brigade to move the duties of a rear guard, and in order to discharge them faithfully every man must be in his place at all times. You are now leaving your comfortable winter quarters to enter upon a stirring campaign, a campaign which you will be filled with blood and fraught with the destinies of our young Confederacy. It's success or failure rests upon the soldiers of the South. They are ready to the emergency. I feel no hesitation to predicting that you, at least, will discharge your duties: and when the struggle does come, that proud banner you bear, placed by the hand of beauty in the keeping of the brave, will ever be found in the thickest of the fray. Fellow soldiers, Texans, let us stand or fall together.

After throwing that handsome bouquet at us the boys, give three rousing cheers for Colonel Hood, took a lingering look at the old camp and rode off in the direction of the Rappahannock. Our tents and cabins were left standing by special orders from headquarters.

The object of that order was to deceive or mislead the enemy who were camped on the Maryland side of the Potomac – opposite.

The artillery and supply trains preceded the infantry and left the narrow roads in a horrible condition. The mud was deep and cold and our progress was necessarily slow. Late in the afternoon we came to "Chapewamsie Run," a deep, swift little stream brimmed of melted snow from the surrounding hills. All the creeks and branches in that country are called runs and some of their Indian names are unpronounceable. If it hand not been for Captain J. T. Hunter, who now lives at Oakwood, Leon County, I would not now remember the Chapewamsie.

As the Colonel was a little worried at the slow progress we were making, he concluded to set an example. Dismounting from his horse he threw the bridle rein over his arm and sang out "Come on, men, right through. It's not deep."

He majestically stalked through the cold ice water which was more than half deep to him, and General Hood was more than six feet tall. Of course, he was well prepared to set such an example, for he wore water-proof boots that came well up on his thighs, so he passed over dry-shod.

With a shivering whoop the two first companies plunged in the cold stream, which came up to the middle of some of the shorties. Pleased with the way the boys took to water, Hood remounted his horse and directed Lieutenant James T. Hunter of Company B who was being Adjutant during the absence of Adjutant R. H. Barsett, who was sick in Richmond at the time, to remain at the ford and see that every man waded the stream. The enthusiasm that actuated the two front companies to follow the example so readily was pretty well oozed out by the time Company B reached the crossing. Lieutenant Jim, mounted high and dry, sat on his horse, telling the men to go right through. Glancing down the stream, I discovered an old mooring spanning the creek, so I made a flank movement in that direction. The Lieutenant saw me and called out, "Go back there, you Company B man." As soon as I knew that he had discovered my motive I started in a run, but he was too quick for me, pausing his horse in the water, he reached the log when I was about half way across it. I tried to avoid him, but he gave me a shove as I attempted to pass, and I lost my balance and landed six feet below where the water was five feet deep. Man, gun, knapsack and accoutrements all went under. Of course that raised a big shout; it was fun for the other boys who had to keep in the middle of the road. I scrambled out of the frigid current and crawled up the slippery bank as wet as a drowned rat and mad as a wet hen.

Lieutenant W. C. Walsh – afterward Captain – was in command of my company and when I returned my place in ranks he and Lieutenant Jim were having high words. They both had their hands on their swords, but other officers interrupted and prevented those two high-strung young chaps from fighting. I was decidedly the wettest, the coldest and the maddest. That little affair created a coldness between the Lieutenant and me – especially on my part, and I had it in for him, as threats went then. I was going to make his life in the regiment a perpetual misery.

"Giles, old boy, get out of there, the Yankees are coming." It was down below dam No. 2 on the Little Warwick, and my post was near. The remark was addressed to me in a kindly voice and I recognized Lieutenant Jim Hunter. He was than an aide on General Hood's staff and while down the river inspecting the picket line discovered that the enemy was attempting to cross the stream, so he hurriedly notified the men to fall back.

After Captain P. P. Porter was killed at the battle of Gaines Mill, Lieutenant Hunter resigned his position as aide and took command of Company A, with the rank of Captain. He led his company at Second Manassas and was desperately wounded there.

For the sake of auld lane syne we took a glass of ice water together at the Bryan reunion. Jim Hunter was a splendid soldier. The incident at Chapewamesie Run occurred early in the war before we volunteers had learned that we could not play in a soldiers backyard or (?) down his rain-barrel. The volunteers of the early days were hard to manage at best and when they were commanded by former playmates it was a trying ordeal for the officers; but we came down to it at last and plumbed the line like old regulars.

It was not long after the Chapewamsie episode until I got into trouble again. We crossed the Rappahannock River at Falmouth, a short distance above Fredericksburg and went into camp on the hills west of the city. During our two weeks' stay there it was drill, drill, drill all the time. They were breaking us in. At dress parade one evening it was announced that a grand division review would be pulled off next day at 2 o'clock.

Now, up to that time I had never seen a grand division review. It was something new and I was anxious to see it and had an idea that a fellow cold see but very little of it from the ranks, so I concluded to play hooky, see the show right and take chances. I found two compatriots who agreed to go with me.

The morning before the review was devoted to cleaning up. We brushed our uniforms, scrubbed and rubbed our guns and had company inspection. When old Colllins, the regimental bugler sounded the call for the regiment to form on the color line preparatory to marching to the parade grounds, Ed. B. Millican with A. Keller and myself started away and march ourselves to the big field, where the grand maneuvers was to take place. We located on a rock fence away back, never dreaming that our regiment would come that way. Many ladies and gentleman from Fredericksburg and surrounding country were there to see the big show. They were all around us on horseback and in carriages, so we felt perfectly secure and happy. The review was a grand sight, as more than 30 regiments were on the field. Batteries were firing salutes, brass bands were playing patriotic airs, battle flags and State flags flashed and waved in the sunlight as regiment after regiment passed in review. Generals, followed by gaily dressed staff officers, rode at the head of their brigades, while Colonels, Majors, Adjutants and couriers pranced past us, each fellow looking as if the success or failure of the Confederacy depended upon him alone. It was s soul-stirring picture, with all the "pomp and circumstances of war, glorious war," attached. We were charmed with the moving panorama before us, and forgetful of all danger until Will Keller slid off the fence like a terrapin, saying as he disappeared behind the rocks, "Look out!" The warning came too late, for just then old Company B swept past us in platoons. The boys say Millican and me perched on the fence and sang out. That attracted Lieutenant Walsh's attention who was marching in front of the first platoon. He shook his sword at us as he passed, but that didn't close the incident. We had though very little of the punishment awaiting us for our willful disobedience of orders until the review was over and we started back to camp. Then we began to discuss the matter very seriously. Keller said he thought we would be court-martialed and shot, as examples were in order. Millican was of the opinion that we would be made to wear a ball and chain for about three weeks. I had an idea that we would be diked out in barrel shirts and labeled "Grand Reviewers" and marched up and down in front of the guard house for the remainder of the war.

There is a great deal of difference between the barrel shirt of 1862 and the shirt-waist of today. The barrel shirt was made of much heavier material than the shirtwaist, and came way below a fellow's belt, holding his arms down by his side, making it impossible for him to salute his superiors. It was a very plain garment; no frills, collar or cuffs, and quite simple in its construction. The head was knocked out of one end of a flour barrel and a hole cut in the other end big enough to slip a man's head in. A dude nowadays can get into a shirtwaist without any outside help, but that was not the case with the barrel shirt. I never heard of a soldier voluntarily putting one on or attempting to do it. Usually it took two, and sometimes half a dozen stout fellows to get the thing out of a refractory soldier. They were not popular in the army although Colonel Van Manning of the Third Arkansas introduced the style in the Texas brigade. While we were in the Shenandoah valley, in October, 1862, I saw six members of that old regiment promenading up and down in front of the guard tent uniformed in barrel shirts and labeled "Straggler," "Forager," etc.

As we sneaked back to camps late in the evening our reflections and conversation were not very pleasant. Some old fool once said that stolen pleasures were the sweetest on earth. I don't believe it. Frank L. Price, Orderly Sergeant of my company, had received his instruction before our arrival, gathered us in and marched us up to the Lieutenant's tent. That individual came out tugging at his baby mustache and looking fierce. We deserved it, and would have been willing to quit at that and promise to be good – but no. After the got through expatiating on duty, obedience, disobedience, discipline, duty and moral behavior he turned to Sergeant Price and said:

"Sergeant, make these men get axes and dig up those stumps," pointing at three old redwood stumps in front of his tent. Soldiers had cut the trees down for firewood and the stumps were about three feet high and 20 inches in diameter, all vigorous, healthy-looking old fellows. The axes being procured, we drew off our coats and prepared for action. Now, if there is anything on earth duller than a Texas campaign after a rot on early spring primates, it is an army ax. We were not happy, although we tried to appear so. We attracted a good deal of attention and the men in camp began to guy us and we could hear voices from nearby and from afar off, saying "How do you like grand division reviews," "Why don't you get a furlough." "Here is your mule, Joe," etc. We tried to at all this, but it was a sickly effort, the kind that a fellow makes when he's not tickled. I had been reared on a farm and was a pretty good woodchopper, but I don't suppose Millican or Keller had ever cut a cord of wood in their lives. While it was generally understood that I was a very lazy boy, there were two things that I dearly loved, and that was a sharp Collins axe and a soft cedar tree, but I was never stuck on army axes or red oak stumps. Just before tattoo the Lieutenant came out to inspect the work. I had my job nearly finished, but Keller had only scratched at his old stump, while Millican had knocked off a little mark and was sitting flat on the ground making the old red oak stump with the back end of his axe. The Lieutenant asked him why he didn't use the edge of his axe, and Millican replied that the back was sharper than the blade. The officer was disgusted and told us to go to our quarters and finish the work in the morning. That morning we received orders to march, and started on our tramp for the peninsula. That was the 7th of April, 1862, and in the January following I visited the old camp after the battle of Fredericksburg.

Fearful fighting had occurred there during that battle and the whole appearance of the country had materially changed. The rock fences were leveled to the ground, the trees bent and broken by shells, and the only things I recognized were the three old mutilated red oak stumps, grim reminders of a grand division review.

Our march from Fredericksburg was without incident, the weather alternating with sleet, snow and rain. General Hood was a brigadier then, having received his commission after leaving the Potomac, pronounced it the severest weather he had ever experienced on a march. We arrived at Yorktown in tolerable condition considering the rain and mud we had to contend with. Here we were assigned the position of First Brigade, First Division, Reserve Corps of the Army of the Potomac. A little later the Federal army assumed the name of the Army of the Potomac; then the Confederates changed to the Army of Northern Virginia, a name Lee's army had until the close of the war. Our first bivouac at Yorktown was about a mile to the rear of defense on the very identical ground occupied by the rebel army of 1781 just before the memorable battle of Yorktown. In some places the old earthworks were still visible, although it had been more than eighty years since Cornwallis surrendered his army to General Washington on the 19th day of October 1781. The (?) took place was at the half way point between the Federal and Confederate picket lines. A monument commemorative of that historic even had been erected there. It has been said of Cornwallis that he positively refused to hand his sword to a common rebel like George Washington and broke the weapon in two by thrusting the blade into the ground and snapping it off at the hilt. Tell a soldier that he cannot go to a certain place and he will immediately want to know the reason why and begin to investigate. We had strict orders not to pass beyond our pickets, but that old monument between the lines had a special attraction. How nice it would be to slip into that forbidden spot and collection a few souvenirs to send home to our friends. Actuated by a silly romance and a disposition to do something that we were not ordered to do, George Robertson, George Nichols and myself concluded to venture in and inspect the old pike. We got there all right, but we did not take time to read the inscription on the monument, and set our faster than we went in. Several other attempts were made by the boys to investigate which invariably brought on a few. Previous to our arrival at Yorktown the enemy had become very aggressive and their sharpshooters had approached to within 150 yards of our fortifications and from two tops and rifle pits they fired at every man who showed his head. Thus they could do with comparative safety, for the troops we relieved were armed with smooth-bore muskets. When the Texans took possession of the breastworks, rifle pits and picket lines things changed very suddenly. We were armed with Springfield, Enfield and Minnie rifles and every man in the brigade knew how to use them. The first day's shoot was a grand success and there was very little bantering after that. Now and then a Yank would get gay and sing out across the line, "Come out of your holes, you rebel rats," but he always took good care to stay in his own. However, quite a number of men were killed and wounded and our stay there was a series of false alarms, night marches, feints, skirmishes and frolics.

Many amusing things happened while we were at Yorktown, and it is better to remember them and tell them here than to relate the sad events and pathetic scenes that occurred there. Fun and misery went with us everywhere, and if we were merry to-day we were sad to-morrow. "But such is life," either in the army or out if it. The little Warwick River separated the two picket lines down at dam No. 1 and dam No. 2. Dam No. 1 had backed the water until the lake or pond was near 200 yards wide above it. On our side the breastworks were built of longs, and the men were quite safe and comfortable after they got behind it. The picket guard was always relieved at night, so as to prevent the enemy from seeing and they practiced the same tactics. Just across the pond, opposite one section of our works, a two-story house had been burned, leaving nothing but the old brick chimneys standing. They stood there in solitude, charred relics of "grim visaged war." One of those energetic Yankee sharpshooters located himself behind the chimney that stood nearest the water and by some means secured a footing that brought his body up to the fireplace in the second story. Then he removed a brick, making a porthole big enough to poke his gun barrel through. That gave him a down pull on our breastworks, and he paralyzed section No. 4, which was immediately in front of him. The boys located him by the smoke from his rifle and blazed away at the old chimney, but it did no good. He appeared to be there to stay. A novel and original idea struck some of the soldiers, and I think it was Sergeant Dick Skinner, who crawled out and hunted up old Captain Riley, who commanded the field battery attached to the Texas brigade. He enlisted the sympathy of that gruff old Irishman, who volunteered to take one of his Napoleon guns down and fix Mr. Yank right. He concealed his cannon in the thick brush a little back from our position and went at his work very deliberately, cutting away the boughs that obstructed his view and sighting his gun with care. While he was getting things arranged to suit him. Mr. Yank was amusing himself by shooting holes through the old hats held a little above the logs on the end of ramrods. Finally old Riley turned his Napoleon guns loose with a roar that rippled the water in the little lake, and filled the woods with black smoke. The shell struck the old chimney center, about six feet from the ground. Brick, mortar and dust filled the air and the lot came tumbling over our way. That Smart Alex of a Yankee came sailing through space, his coat tails standing in the breeze like the latter end of a comet. Arms spread out wing fashion, hatless and without a guan, he fell in the mud, but fell running. Wet as a beaver, he scrambled up the sleek bank and was exposed to our fire for more than 50 yards. It was a run for life, with whoops, yells, bullets and laughter following him. We could hear the Yankees shouting and hollering "Go it," while the "Johnny Rebs" took at least 50 wing shots at him, but he made good his escape. Old Captain Riley was so elated with that fine shot he made and the fun it created that he laughed until the tears rolled down his wrinkled old cheeks.

The South was full of fire-eaters when the war first began. They raved and ranted and swore than the Yankees wouldn't fight and that one Southern man could stampede a whole cava yard of them. Very few of those fellows went to the front and what few did go were like bullets who rode through saloons and shot out lamps before the war were not worth a flip when the final tug came. There was a big, red faced fellow in my regiment who was a chronic growler, fault-finder and blower. The army was too slow for him. The officers didn't know their business and the Yanks were all cowards. He was so boisterous and dangerous that the boys gave him the name of "Maneater," as he often expressed himself as wanting two or three raw Yanks before breakfast to give him a good appetite for hardtack and bacon. One day Maneater was detailed for picket and sat down below dam No. 2. There were no breastworks down there and the men had to take shelter behind trees, or stand out in the open. He was stationed near a big old scaley-bark hickory, but he scorned the idea of taking refuge behind it until a Yankee sharpshooter clipped a twig close to his left ear, then he skipped for the scaley -bark. He shook his fist in the direction of the enemy and dared them to come out and fight like men. Bill Calhoun was a few yards left of Maneater, sprawled out on the ground behind a fallen tree. Bill said Maneater was such a d__d fool and made so much noise that the Yanks soon located him and marked him for their own and began chipping off the bark on both sides of Maneater's tree. As Bill told it "Maneater" suddenly dropped his gun, grabbed his jaw with both hands and began whizzing around like a top, yelling "Lordy! oh Lordy!" then flopped on the ground.

Bill thought he was killed or mortally wounded, so he crawled to him, and laying his hand gently on "Maneater's" back said: "Poor old fell, are you hurt much?"

"Hurted, did you say?" Hell and damnation! Jaw busted wide open!" yelled Maneater.

Bill managed to get the dying man back behind a little knoll where they were safe from the enemy's fire.

"Let me see the wound,' said Bill, after assuring Maneater that they were out of range. Maneater slowly removed his hands from his jaw and Bill broke out into a loud laugh. Maneater swore and called Bill a heartless wretch and many things much worse. A bullet had knocked off a sliver of bark close to Maneater's jaw and a few fragments had scraped his cheek, producing a drop or two of blood. Bill told it when en came to camp, and of course made it worse than it really was. The boys deviled poor Maneater nearly to death after that, and when a group of them saw him approaching or passing they would grab their jaws with both hands and to waltzing around shouting, "Lordy, oh Lordy!" "Hurted, did you say?" Jaw busted wide open!" etc.

Maneater tried to get furlough; he tried to be promoted to mule driver; he tried to get a transfer, but he had no additional pull and no influential friends, so he finally gave it up, quit wanting raw Yankees for appetizers and made a pretty good soldier.

General Johnston began his retreat from Yorktown on May 3, 1862, but it was 4 o'clock on the morning of the 4th when we silently abandoned our line. The Texas brigade was the last infantry to leave Yorktown, and we had not gone two miles when we heard a tremendous roar behind us. The entire sky was lighted up with a red glare, the decayed part of Yorktown was on fire. The blowing up of the old town was all a mistake caused by the inquisitiveness of a few cavalry vidette left in the rear to watch the enemy. Several mines have been planted in several of the houses and in the streets to blow up the Yanks when they entered the town, ignorant of the fact, these enterprising troopers burst open a door looking for whisky, and though unsuccessful in their search, they came out considerably elevated. The stepping on a fuse by one of these gay cavaliers started the whole shooting match, and soon it seemed as though a fierce battle was raging in the ancient little city. This premature explosion notified the enemy that the Confederates were in full retreat and they came in legions. Every few miles we were thrown in line of battle, but the vanguard that pursued us were shy, and we passed through Williamsburg before night and bivouacked two miles west of that historic little burg. We passed the main army about a mile east of Williamsburg, where it had been halted.

Hundreds of men were busy felling trees and digging rifle pits preparatory to receiving the enemy. The Battle of Williamsburg was fought the next day, May 5, 1862, and was a stubbornly contested affair, but the enemy was checked.

Williamsburg was founded in1623 and is the oldest incorporated town in Virginia. William and Mary's College looked grim and old as we passed it in retreat. With the exception of Harvard it is the oldest college in the United States.

Three of our Presidents – Jefferson, Madison and Monroe – were educated there and back in the early days Williamsburg was the Mecca of Colonial splendor, when George III was King of the realm and every gentleman wore a sword. It was the Colonial and State Capital until 1779, where the dames, the belles, and the gallants of the Colonies assembled to intrigue, make love, flirt and trop on the light fantastic toe to the tune of the spinet or the merry strains fo the old fiddle.

Where the Peggies and the Pollies,
The Marthas and the Dollies,
With their dainty satin slippers – just the thing –
Stepped lightly through the minuet
With George and Patrick in it.
When the blush was on the roses in the spring.

Val C. Giles, Austin, Tex.

Sergeant Val C. Giles
4th Texas Infantry

Battle Of Eltham's Landing

History has said but little about the battle of Eltham's Landing, fought by the Texas Brigade on May 7th, 1862.

The result of that sharp, quick engagement was worth a great deal more to the Army of Northern Virginia at that time than the great Confederate victory at Fredericksburg, where Lee whipped Burnside to a finish in December 1862. At Eltham's Landing General Franklin was checked and driven back to his gunboats and transports, and the way cleared for General Johnston's retreating army.

As soon as it was known by the commander of the Army of the Potomac that General Joseph E. Johnston was withdrawing his troops from Yorktown and falling back toward Richmond, early in May, 1862, General Franklin of the Federal Army was hurried up the York River on transports with 15,000 men to intercept and cut off General Johnston's retreat.

General Johnston halted the main part of his army at Williamsburg, and fought McClellan there on the 5th of May, 1862.

The Texas Brigade was the last infantry to cut out of old Yorktown at four o'clock on the morning of the 4th, bringing up the rear of the retreating army.

To be at the tail end of a retreating army is regarded in warfare as the "post of honor." It was an honor (so-called) that often fell to Hood's old Texas Brigade, and is decidedly the meanest and most hazardous position in a retreating army.

Every little while we were thrown into line of battle, for the enemy was pressing us hot and heavy in overwhelming numbers.

We marched through the historic old town, and a few miles beyond turned to the right, leaving the Richmond and Williamsburg road, continuing the march until 10 o'clock at night.

I say we marched through the historic old town but I might add, to my discredit as a soldier, that I played hooky and did some straggling when my regiment reached the town.

Wm. F. Ford, Garland Colvin and myself quickly dropped out of line and hid behind an old red brick church until the brigade had passed.

We then sallied forth to inspect William and Mary college, where two presidents and Patrick Henry got their inspiration for American independence, the very thing that we were then struggling for. From the time Pocahontas rescued one of the Smith family from the tomahawks of old Powhatan's braves, that part of Virginia lying between the James and York Rivers, known as the Peninsula, had been historic and romantic section of the Old Dominion.

Back in colonial days the belles and swells gathered in old Williamsburg, at the time the capital of Virginia, and danced the minuets to the music of spinet and fiddle.

We inspected several old churches built of red brick brought from England in sailing vessels before the inhabitants knew that they could make brick out of Virginia clay.

The streets were thronged with artillery, supply trains and soldiers, and the huckster was there crying his wares, for war had no terror for the Virginia huckster with his old poor horse and two-wheel cart.

After taking a hurried view of the exciting scenes of a retreating army and filling our haversacks with black ginger cakes, we lit out on the trail of our old brigade.

It was twelve o'clock before we reached camp that night, and found everybody asleep except the camp guard.

The guard let us pass, and we were never punished for our disobedience of orders, although we were marked absent without leave at roll call that evening.

On the morning of the 5th we were up and going by sun up. We heard the "opening roar" of the battle of Williamsburg, but pressed steadily on leaving the main army to fight it out. Of course we didn't know where we were going, or why we were being withdrawn from the main army on the eve of battle. We moved on through a drizzling rain, deep mud and narrow, crooked country roads, and when we went into camp at night, we were many miles from the battlefield of Williamsburg, where General Johnston completely checked the vanguard of McClellan's army.

On the 6th we made another long hike through mud and rain, and the dim old roads we traveled were so badly cut up and narrow that part of our provision train failed to reach us until the next morning.

The wagons hauling our flour and bacon came in about ten o'clock at night, but those loaded with cooking utensils were stuck in the mud many miles behind. The country where we camped that night was densely wooded with red oak, pine, sugar maple, and beech. It was in the spring and sap was up, so the boys peeled off the back of many of the trees and made trays to knead their dough. Then they wrapped the dough around their ramrods and cooked it before the fire and broiled their bacon on the coals. Maybe that's why Confederate soldiers were called "doughboys."

An epicure would shudder today at such grub, but when a fellow is in the army, young, healthy and hungry, it is the finest eating in the world.

On the morning of the 7th, we started early, still bearing to the right. It was raining a slow, misting drizzle, and as we plodded along all muffled up, trying to keep dry, the 4th Regiment in front of the Brigade, and Company A in front of the 4th, the first thing we knew it was bang, bang – we had run into the Yankee pickets unexpectedly with guns empty.

Hood rode at the head of the column, and the pickets fired at him, but failed to hit him. One of the Minie' balls intended for Hood struck Corporal Sapp of Company H in the head, inflicting a severe wound. There was but one man in the regiment that had his gun loaded, and that was John Deal of Company A.

John dropped to his knee and killed one of those pickets dead in the old road.

Hood afterwards complimented John on the fine shot he made, but chided him very gently for loading his gun without orders. The fact was, that the evening before John discovered a bunch of fat porkers in a woodlot near the road. He concluded to confiscate one of those shoats for the good of the cause, but just about the time that he drew a bead on a nice fat fellow, the owner of the hogs showed up and advised the men to move on, and John very promptly moved.

As soon as the pickets fired on the head of the column, Hood galloped back hurriedly, calling out: "Into line on first company and load."

The regiment rushed forward and formed on the crest of a hill overlooking the York Valley. Two platoons from Company B, under the command of First Lieutenant W. C. Walsh, were ordered forward as skirmishers. We deployed and started down the hill across from an old field at a double-quick pace, closely followed by Company G, under Captain J. W. Hutcheson.

We drove the enemy back to the woods, and the balance of the brigade soon came up.

We were again ordered to advance and "feel the enemy." That "feel the enemy" was a favorite order of Hood's and a very hazardous and unpleasant duty to perform.

General Franklin had landed his transports and disembarked a large body of infantry. The woods were dense and we worked our way through the underbrush and vines until we ran up against a line of Federal infantry. When they began firing on us, Lieutenant Walsh gave the order "Lie Down," and every fellow dropped to the ground and the bullets passed over us, but did some damage and the bullets passed over us, but did some damage to the main body of our troops behind us. The battle was now on, and B and C companies between the two firing lines. The Federal gunboats in York River opened with grapeshot and canister, making the old woods howl. We were now in a very unpleasant position between the two firing lines, and realizing this fact, Walsh shouted, "By the right flanks," but by that time our main line came up and the enemy was falling back toward the river. They made a second stand, and for twenty minutes that was one of the loudest little fights I was in during the war. It may be that I thought it was so noisy because it was my first real battle. We had experienced a good deal of unpleasant picket fighting at Dam No. 1 and No. 2 on the Little Warwick River, while at Yorktown, but the battle of Eltham's Landing was our initiatory battle. It was a mere skirmish compared to some we were later on.

Nevertheless, Companies B and G of the 4th Regiment went down in that wilderness ahead of the brigade and brought on the fight. We never had a man killed.

If my memory serves me right, the First Texas had twenty-eight men killed and wounded. Lieutenant Colonel H. H. Black was mortally wounded, and Captain Decater and ten men were killed on the field. Captain Denney, Commissary of the 5th Texas was also killed early in the action. The Federal report shows that their loss was more than two hundred men.

Just before sundown we were withdrawn from the dismal woods and went into camp back on the hills for a few miles from the battlefield. That night I was detailed for camp guard. A strong picket line had been established between our camp and the enemy. I had been on post only a short time when an ambulance came slowly up the narrow muddy road. I halted the drive and asked him where he was going.

He replied, "I am trying to find the field hospital. I have a wounded officer in my ambulance."

I asked him who it was, and he said, I don't know."

I took his lantern and looked in the ambulance, where I saw a slight, delicate-looking man with a gray blanket thrown over his lower limbs. His head was resting on a knapsack, his eyes were closed and his face very pale. He muttered something I did not understand. I handed the lantern back to the driver and told him to move on. It was Lieutenant Colonel H. H. Black of the First Texas Regiment, who had been mortally wounded in the battle. He died in the ambulance that night and was buried in the piney woods, near the old road. I don't suppose there is a soldier living today that could point out that lonely grave.

Next morning, May the 8th, we struck in the direction of Richmond. The narrow country roads that we had been traveling were bad enough, but when we struck the Richmond and Williamsburg road in the wake of thousands of Johnston's men, it was simply indescribable.

We had to wade through soft mud up to our knees, and some of the language used by the soldiers on that occasion would not look well in print. At one place, after passing through a sea of yellow mud, we were ordered to halt and rest. That was the first time I ever saw General Joseph E. Johnston. General Hood was sitting on his horse nearby, talking to some officers, when General Johnston, with his escort rode up.

He greeted Hood pleasantly and said, "General, I sent you up to West Point to check the enemy, but not to bring on a general engagement."

"Well, General," said Hood, "those people were so saucy that the boys wanted to give them a little thrashing, and I let them do it."

A few more remarks were passed between the two officers, then General Johnston rode on in the direction of Richmond.

General Joseph E. Johnston

(Library of Congress)

I remember just how he looked on that occasion. He wore a gray cloak with red trimming, the left half of his cloak thrown over his shoulder. His uniform coat was buttoned up to his chin, and a little red rose was pinned on his left breast, and he looked every inch the soldier he was.

When we reached the Chickahominy River, it was eleven o'clock at night and dark as pitch. A miserable old rickety bridge spanned the slugging stream. By this time the mud in the narrow road had been worked up to a thinness that would have floated a canoe. The men were mad, tired, hungry, and in no humor either to receive or to obey orders.

General Whiting, who commanded my division at the time, had stationed himself on the east side of the old bridge, and as the men came up through the mud and slush he kept singing out, "Close up, men. Move lively." Some fellow in the ranks called out, "Oh, close your bread-trap, and give us a rest."

The general shouted angrily: "Do you know who you are talking to? This is General Whiting."

"General Whiting be damned. You are nothing but some long-tongued courier," came from the dark, muddy road.

"Officers, arrest that man," yelled the General, but nobody paid any attention to the order, and with an oath the general rode across the bridge and disappeared in the darkness.[6]

6. Laswell, Mary, (ed.), *Rags and Hope, The Memoirs of Val C. Giles, Four Years with Hood's Brigade, Fourth Texas Infantry, 1861-1865*, Coward-McCann: New York, 1961, 92-99.

Sergeant Samuel T. "S. T." Owen
4th Texas Infantry

State of Va. May 22, 1862

Dear Mother,

I this morning take my pen in hand to inform you that I am well and doing well and when these lines come to hand I hope that they will find you all well and doing well. Well Mother I have understand that father have left you at home by yourself but mother I hope that it is al for the better.

I hope that he has left you plenty to eat til he can get back from the war I think that the gretest fight that ever has ben none I think that it wil be hear at Richmond We hav a grate many men hear but I hav not told you nothing about our little fight the other day on the 7th of May our regiment was 2 wonded nary one kild there loss was grae it last about 2 ours We had some mity hard marches but I think we air station now for the Yankees the recruits got in the others day We was laing in the brsh looking for the enemy every minit but they did not come the recruits. Air al not Well billy clannyham has the measles and several of them has the dierra the old company air all well I want you to write me as often as you can so nothing more at present only remains your effection son until death.

S. T. Owen[7]

J. B. Polley was wounded at the Battle of Darby Town Road on October 7, 1864 and was wounded in the lower right leg and lost his right foot. In 1910 he wrote the first official history of Hood's Texas Brigade in a book titled, Hood's Texas Brigade: It's Marches, It's Battles, Its' Achievements.

---

7. Letter, Samuel T. Owen to Mother, May 22, 1862, Hood's Texas Brigade files. Texas Heritage Museum-Historical Research Center.

3rd Corporal J. B. Polley
4th Texas Infantry

Around Yorktown
Camp Near Richmond, Va.,
May 19, 1862.

3rd Corporal J. B. Polley, 4th Texas Infantry.

(Texas Heritage Museum Historical Research Center)

Arrived at Yorktown, we camped a mile and a half to the rear and right of that dilapidated old town. It was here, you know, that Cornwallis surrendered. The embankments thrown up during the Revolutionary War are yet in a fair state of preservation, and would likely to have been exceedingly interesting to me had not the present war - in the shape and terror of bombs from a Federal battery - furnished a more practical and exigent subject for reflection. Some of my comrades grew quite enthusiastic over the fact that we were on historic ground, made sacred by Washington's great victory, and eloquently insisted that the scene should inspire us with extra courage and patriotism. Suspecting that the larger part of their enthusiasm originated in the canteen of whisky they bought from a blockade runner, I tasted it, but it aroused no corresponding sentiments in my pacific bosom.

About two o'clock on the morning of May 4th the pleasing information was communicated to the Texas Brigade that to it had been granted the proud distinction of serving as the rear-guard of the Confederate army in a retreat from the peninsula. In fact, all the other troops of our army had already folded their tents, and without giving us the slightest hint of their intention, had hours ago marched away toward Richmond, and under these circumstances even the compliment paid the brigade by giving it the post of honor failed to relieve us of a feeling of lonesomeness and insecurity. Just as day appeared we took up the line of march, and anxious to put as much ground as possible between us and the presumably fast following Yankees, stepped out in our very liveliest manner. But either because they knew that Texans were the rear-guard, and feared to attack such desperadoes, or were not fleet enough of foot to voer take us, we went peacefully on our way and overtook the main body of our army about four miles from the old colonial town of Williamsburg – with the proud and inspiring consciousness thrilling our patriotic bosoms of duty well performed by heroic efforts to get beyond reach of a dastardly enemy. Yet although terribly tired by the rapid march over eight miles of the muddiest road imaginable, we halted not an instant, and leaving Williamsburg to our left, hurried rapidly on toward York River. We were in good luck, for after an hour or two of tramping, the roar of artillery and the roll of musketry – fortunately many miles behind us – smote upon our unaccustomed ears, and we could reflect exultingly, that while the honor and glory of having been the rear-guard was ours beyond dispute, we had yet escaped all the dangers of that distinguished position. General Hood neither halted, changed the course of the march, nor furnished us with a single particular as to his intentions, but hastened us on with a speed which appeared to indicate so strong a desire to reach a haven of complete safety that the proceeding met our most hearty approval and cooperation; not a man straggled – not one lagged in the rear. As a result of this unanimity of purpose we were soon beyond recall and sound of the battle and made camp that night in the heavy timber about four miles from Eltham's Landing.

Here we remained until the 7th, when at dawn we advanced nearer to the landing and the enemy. General Hood and his staff were a hundred yards in advance of the Fourth, Company F next to the leading company, and we were approaching a large deserted house situated on an eminence overlooking the wide valley of York River.

Between us and the mansion were some cavalry pickets, who like veritable dummies, had sat on their horses and permitted a company of Yankee infantry to shelter itself behind the building. Hood reached the picket post scarcely a hundred yards from the house, and immediately a squad of bluecoats stepped out in plain view and poured a volley into us – doing no greater damage, however, than to give us a terrible scare. We had been marching at will and in column, and except that of John Deal of Company A, not a gun was loaded. It was a complete surprise. We were in a newly cleared field full of pine stumps, and, with the instinct of self-preservation suddenly aroused, every man except Deal – who immediately knelt, fired and mortally wounded the sergeant of the attacking force – hastily sought the protection of a stump, loading his gun as he ran. Hood came dashing back, shouting to the regiment to fall into line, and as every stump I made for was appropriated by a quicker man, and I managed to load my gun, no option was left for me but to be among the first to obey orders and place myself in approved battle array.

Not half a minute elapsed, though, before every man of the regiment was in ranks, and then came the order to charge. Rushing bravely and furiously to the crest of the eminence, we were overjoyed at seeing the enemy fleeing across an open field to a skirt of timber half a mile away – their precipitate flight availing them little, however, for not a man of the fifty or more in sight and range escaped wounding or death.

To the right of the house grew heavy timber, and there we had deployed into skirmish line a number of Yankees were killed and captured. After a while the brigade moved forward across the field into the woods beyond, but the enemy was driven back so rapidly by our skirmishers that not a single one capable of doing duty came within my view. I made no complaint, and as long as I kept out of their sight was thoroughly content. The other two Texas regiments – the First and the Fifth – had hot fights which they won by gallant charges, and in two hours or so the Yankees were forced to take refuge in transports protected by gunboats, which shelled the woods until night.

Thus, Charming Nellie, began and ended your friend's first experience under fire. He did not distinguish himself, it is true, but he finds great consolation in the fact that neither did the enemy nor the Virginia cavalry, who, by their carelessness, almost caused the Fourth Texas to show "the white feather" in its first engagement. Here I looked for the first time on the dead and wounded of a battle. After the fighting was over, Jack Sutherland and I went to a poor fellow who was mortally wounded, and filling his canteen with water did what else we could to make him comfortable. He admitted being from Wisconsin, but absolutely refused to name the particular command to which he belonged, saying it was against orders. He was just about my age, and it was not a pleasant thought that someday soon I might, like him, be mortally wounded and left in the hands of the enemy. I do not often indulge in such grim fancies, but in his presence could not avoid them.

Three days' rations were issued to us the day before we left Yorktown, and on the morning of the 8th, our haversacks being again empty, four ears of corn were dealt out to each man. When parched it was not at all bad to eat to hungry soldiers, and we soon became genuine Cornfeds.[8]

8. Polley, J. B., *A soldier's letters to Charming Nellie*, New York, NY: The Neale Publishing Company, 1908, 32-36.

3rd Corporal J. B. Polley
4th Texas Infantry

As we were no longer the rear guard we travelled fast and by three o'clock in the evening had made full 30 miles and encamped within three miles of West Point (Eltham's Landing) where the enemy were landing. Here we cleaned our guns thoroughly and here too our rations began to run short. I had plenty to last me three days longer but had to divide with those who had been less provident and thus found myself on Tuesday morning minus bread. Capt. Owens our commissary was about to issue one day's rations of flour but was ordered not to do so as he expected an engagement every minute. About 3 o'clock P. M. we were called out on the color line where we had to lie until night. While here our rations of hard bread and bacon was issued and was quite a trial to many who had little breakfast and no dinner. After dark we joined the Brigade. After some time spent here the 1st Texas was ordered to a point nearer West Point to act as Picket Regiment. The other three regiments moved some half a mile back and encamped. Next morning by sun up we were under way towards West Point the 4th Texas in advance. Gen Hood and Staff were about 200 yards in advance of us. Just as we had gained an open ridge over-looking the York River Valley, and as our Gen had halted near a mint house, a courier came past shouting that the Yankees were just ahead of us. 50 yards further on and some 20 bullets came whistling over our heads fired by some Yankees behind the house and at Hood who was however unharmed. Not a gun in the Reg'm't was loaded and many of the men in advance dropped behind stump to load. Our gallant young Major galloped back by us shouting "load load" Gen. Hood ran up and springing from his horse sang out "Form a line here my old 4th" Seeing that there was considerable confusion among the men he cried, "can't my old Regiment form into line of battle?" The men gathered around him the Major galloped in front of us and shouted "come on boys" The men gave a yell that made the woods ring and sing after him very few of them having loaded their guns and ran to the brow of the hill. One Yankee was killed on the hill four others just about disappearing in the edge of the woods 400 yards distant were tumbled over in fine style. One man who was running through the open field to our right was shot and turned a complete summersault. The men had halted on up the hill and just here Marshall, whom no one had seen or heard during the affair, rode up, his overcoat under his horse's bell his cap in hand and himself drawn up in his saddle. Co. "A" was ordered to deploy as skirmishers into the wood on our right and the Reg't to lie down. We had to move nearer the wood and just as we had sat down again Dick Skinner of our company killed a bold Yankee who had crept up within 25 yards of us. Several of the boys had taken aim at him but were deterred by others crying out that he was one of Co. "A." Co. "F" was immediately ordered to deploy as skirmishers. We advanced into the wood, Co. "A" giving us room. Our men did not get a single shot while here, Co. "A" having forced the enemy far to our right. Co. "A" were shooting constantly and killed some five or six, altogether ten or twelve were killed on and from the hill and in the woods. The Regiment was now marched to the edge of the woods where Co. "A" and "F" rejoined it and where Co's "B" and "G" of the left wing were sent forward to clear the way. They found the woods alive with Yankees. As we advanced we could see dead bodies, knapsacks, canteens, blankets and overcoats lying on all sides and could hear the moans of the wounded. Our skirmishers seldom missed and made many a Yankee bite the dust.

In this manner we advanced nearly a mile into the woods and finally were drawn up into line of battle to act as reserve for the skirmishers. We laid flat down in an old ditch and while here the 1st Texas who had been brought in to support us met a large body of the enemy by our left-flank. The latter marched up in fine style near where the 1st was crouched in a ditch. When about 50 yards distant when Col. Rainey shouted "rise and give them hell boys". They gave the Yankees one effective volley and then charged them, putting them to complete route. Lieut. Colonel H. H. Black as brave and gallant a man as ever lived was mortally wounded by the enemy's fire. Capt. H. E. Decatur was shot dead and some 15 Texas killed and many wounded. But their loss was nobly avenged. Nearly 100 of the enemy being killed and wounded. The 5th Texas had entered the wood by another route and had a severe little fight to our right. The 18th Georgia Regiment had been ordered to guard a battery stationed on the hill where Hood was first fired upon and thus were not in the fight at all. They would have done brave fighting, I am confident, had they had the opportunity. From our position behind our skirmishers after they had been called in we about faced and went over to support the 1st Texas laying down in a ditch immediately behind them. Since writing I have learned that Hampton's men did not meet the enemy at all, their only fire being by mistake on the 5th Texas.

While here Hampton's Legion filed past and engaged the enemy farther in the wood near the 5th Texas on whom they fired once by mistake killing and wounding two or three Texans. Capt. W.D. Denny the commissary of the 5th was killed sometime during the engagement. In all from Regiment some six were killed and many more wounded. Hood did not understand the position of the enemy or our Reg'm't would have had a better chance to distinguish itself. Two Regiments of Anderson's Tennessee Brigade came in to support Hampton but did not meet the enemy. While we were supporting the 1st the enemy were throwing shells over us into the wood. Few could avoid dodging as the dangerous missiles whizzed over our heads sometimes bursting just above us. Just as Anderson's Regiment had passed, Hood received a message from Whiting how had entered the woods by the main road to our right to the effect that a Division of the enemy were advancing in order to turn our left flank. We immediately marched out of the wood upon the hill. Here hood learned that there was no more fighting on hand and so marched us back to our camp of the night previous. Gen. Johnston, I have learned considers that this fight saved his army. While Co's "B" and "G" were skirmishing our gallant young major distinguished himself. He was constantly riding back and forth behind our men encouraging them and while doing so saw three of the enemy in a bunch. He charged upon them and presenting his six shooter told them to drop their guns. This two of them did the third one springing behind a tree. The Major turned his pistol to bear upon and told him "show your head damn you and I'll shoot it off." The fellow kept dark though until two of our men came up and took three prisoners. In all, by our Brigade, some 50 prisoners were taken here and between 4 and 500 killed. The wounded Yankees expressed great surprise at our shooting so well at such long distances and informed us that we had been fighting the sharpshooters of their army those whom our men had fought at the dams near Yorktown. In our Regiment but two or three were wounded flesh wounds, at that none were hurt of the Mustang Grey's. The evening after our skirmish we had corn issued to us by Gen Hood. All were willing and glad to take it as our rations were not plenty. That night, as Hood expected the enemy to advance, we laid on our arms in an open wheat-field until just before day, when with the utmost silence we took up the line of march for New Kent Courthouse two miles this side of which we encamped. Here Col Black of the 1st Texas was buried with no mark over his grave save the initials of his name and the word "Texas" in large letters cut with my knife on a post near his grave. About noon on the following day we marched some ten miles and were halted nearly two hours to await the result of a cavalry fight behind us and to support our cavalry if necessary, then marched on to a ground some half a mile distant from here, where we stopped until yesterday having camped one night there Yesterday evening we formed in an old field there, laid up our arms till dark, and then moved over here. We are now I learned awaiting the advance of the enemy and are the extreme Infantry rear guard. From all I can learn our Generals hope to draw the enemy into a general engagement here so as to deter their advance and give time for the completion of our works around Richmond which is only 20 miles distant. Should they face us we will have 100,000 men here that many being only a few miles in advance of us. A grand battle must assuredly come off here within ten days. We have got the enemy off of the water and will assuredly whip him as we have always done. Not that advantage I don't think our General intends McClellan time to entrench or fortify and as both sides seem to have staked their all upon this last effort, and as our men have become more desperate and determined than ever we must and will whip them. Since our skirmish of the other day our men who were before eager for a battle are now perfectly willing could they do so honorably to go home without further fight. I'll say that they have seen more fun at practice than in the fight.

Finished Sunday May the 17th 1862 while sitting under a pine tree in a wood 20 miles from Richmond.[9]

3rd Corporal J. B. Polley
4th Texas Infantry

THE DAILY EXPRESS
JULY 15, 1906

These reflections are inspired by a roster that lies before me. It is that of Company H of the Fourth Texas Regiment of Hood's Brigade, carefully compiled and corrected by surviving members of the company. This company of Confederates was organized in Grimes County, Texas, on the 7th day of May, 1861. Just a year later, the Fourth Texas received its first baptism of battle, the first man of the brigade to shed his blood being a member of Company H by the name of H. T. Sapp. He is now living in Harris County, Texas, but was so grievously wounded at Eltham's Landing on May 7, 1862, that he was discharged from the service. He did not remain discharged, however. No sooner had his wound healed than he returned to his old command and re-enlisted, to serve faithfully and with great gallantry during the war and until the surrender of Lee at Appomattox.

---

9. J. B. Polley, Diary, Unpublished manuscript, Hood's Texas Brigade files, Texas Heritage Museum-Historical Research Center, 75-83.

3rd Corporal J. B. Polley
4th Texas Infantry

THE DAILY EXPRESS
MAY 2, 1909

Historical Reminiscences

Conducted by J. B. Polley, Floresville, Texas.

The following "o'er true tale" is forward by a comrade at Galveston, who lest his own name lead to the identification of the hero of the story, asks that it shall not be given. Acquainted with all the facts as we our self are, we not only vouch for the absolute truth of the story, but commend the discretion manifested by its writer in concealing his identity from the general public. It is best he should, for while the survivors of Hood's Texas Brigade are too honest to deny that among them were a few comrades to whom the smell of burning powder was weakening, and the angry hiss of a missile thrown from the mouth of a weapon in the hands of the enemy, horrible in proportion to its size and contiguity, they neither boast of it themselves nor like that it should be proclaimed from the housetop. However, they will not object to our correspondent's expose of an incident over which thy have so often laughed.

"A regimental commissary's duties," writes our correspondent "did not, during the Civil War, require his presence on the firing line, and especially when that was enveloped by smoke and noisy projectiles. On the contrary, they required him to stay as far from it as he could while remaining close enough to furnish the fighting contingent with rations as needed. And to make sure of supplying the demand he must often be riding far in the rear, for the herds of cattle that furnished the meat and the mills that supply the breadstuffs for an army are seldom found grazing, or lying around loose, in the near neighborhood of where two armies are having 'a small bickering.'

"The fact that one is a commissary, though, does not prove him a coward, and therefore the standing order which prohibited either a commissary or a quartermaster from exposing his life to the danger of being 'snuffed out' by musket or cannon ball or by fragment of a shell. But to those of either class who served with the Texians in Virginia, a more effectual warning against giving way to the impetuosity of their valor came from the sad fate of three commissaries, who in disobedience to orders had accompanied their commands and fallen in battle. "The first of these was Captain Denny of the Fifth Texas. Although possessed on May 6, 1862, of so strong a premonition of death the next day that he spent half the night writing farewell letters to relatives and friends he insisted so pleadingly of being allowed to accompany his regiment into battle at Eltham's Landing on May 7, that Colonel Archer then commander of the Fifth, finally gave his reluctant consent. While riding with Colonel Archer at the head of the regiment they approached the brow of a hill that overlooked the valley of York River. Supposing that Confederate cavalry outposts were yet in advance of them nobody suspected attack. It came, nevertheless. As the head of the regiment came within a hundred yards of a log cabin, standing on the crest of the hill, a dozen or more Federal pickets stepped from behind and fired straight down the advancing line. Two or three men were wounded and one killed, and he was Captain Denny.

"The second commissary killed was Capt. Tom Owens of the Fourth Texas. Unlike Captain Denny, no premonition of death was vouchsafed to him, but like Captain Denny, he begged long and hard for permission from Col. John Marshall to go into the battle of Gaines' Mill with the boys he had persuaded to enlist in the Confederate service. He had promised their mothers, he said, to stay by them in the fight, and rather than be false to his trust he would resign his commission and go in as a private. His request was granted, and an hour later, when the Fourth Texas came in plain view of the enemy's infantry and artillery and breasted the storm of death which swept the crest of the hill on which they began their memorable charge on their strong lines of breastworks, Capt. Tom Owens fell, mortally wounded.

"Then came the sad and untimely death of Maj. Tom Hamilton, brigade commissary. At what battle this occurred I am unable to state, but it was in one of the hardest of many hard contested battles by Hood's Texas Brigade. He was a young man, but a few years past the twenty-first milestone of life, and had been taken from the ranks to serve in the responsible position of brigade commissary. He, too, sought and obtained permission to go into the battle with the boys, and he, too, was killed.

Private B. L. Aycock
4th Texas Infantry

After two months in training camp near Richmond, early in the fall of 1861 (November 20), the brigade was ordered to join the Army of Northern Virginia, then facing the Union army at Centerville. It was given out that an engagement was imminent, but this turned out to be a false alarm after we reached the line on the Potomac. We marched a great part of the way, and our position was on the right wing of the army at Dumfries, some thirty miles down the line from Centerville.

Here we went into winter quarters and saw little but the prosy camp life all that winter of 1861-62. We were then under Gen. Joe Johnston. In March (9th) we took up march to Yorktown, quite a change of base, where the Federal General McClellan expected to either capture the rebel army cooped up in the peninsula, or compel its retreat to the Chickahominy River, a stream bent around Richmond, some seven miles from the Confederate capital.

On May 5, 1862, the retreat of our army from about Yorktown began, and here the Texas brigade was given — as the word came to the ranks — the post of honor — that is to say, the post of danger — to be the rear guard of the army. But, after an all day's march, as we passed through Williamsburg, another command took our place as rear guard, and before any sign of pursuit by the enemy as far as we could see. Several miles after this change, that same evening, the enemy overtook and attacked the force left behind, and here the bloody battle of Williamsburg took place. Thus we escaped, unintentionally, a trial of arms with the enemy. As we were still in the peninsula, we were hurried forward to meet an expected attack at Eltham's Landing, where the enemy did disembark from his gunboats, and a small engagement took place. Instead of cutting off our retreat, they were too late. There I saw the first bloodshed in our brigade. After one day here, the Federals betook themselves to their boats, and our march was continued to the north side of the Chickahominy.[10]

---

10. B. L. Aycock, "The Lone Star Rifles," in *Confederate Veteran*, February 1923, Volume 30. No. 2, 50.

Private Wintul E. Barry
4th Texas Infantry

Daily Express
October 20, 1907

So intent was our squad of fifty (at Sharpsburg) in dealing death and destruction to the enemy in our immediate front, that before any of us realized the possibility of it, the First Texas, their ammunition exhausted, and but fifteen or twenty of them left, fell back from the cornfield, and the enemy pushed forward and got into our rear. It was a nasty predicament to be in, and to escape it, three or four of us made a break down the lane, leaving as we thought, the most of our comrades in the lane wounded or dead. But we had gone, but a hundred yards when we ran into a New York regiment - the Thirty-third, I think

They were lying down as we came up, but no sooner saw us close to them than they rose and poured a volley into us, killing two of my comrades, but wonderful to state, never touching me. Of course, I surrendered; what else could I do? For a while, though, after I did, I wished I had not, for when it was discovered that I was Texan, the New Yorkers got in a terrible rage. They had belonged, they said, to Franklin's division which we had encountered and repulsed at Eltham's Landing, near West Point, Va., on the 7th of May preceding, and they insisted that on that occasion we had not only fought them with negroes but had also cut the throats of the wounded Federals who had fallen into our hands.

Indignant and truthful as was my denial of both charges, it was of no avail, and to punish me they forced me, over my solemn and energetic protest, to take a place in front of their line and join in the attack they immediately made.[11]

---

11. William E. Barry aka Wintul E. Berry? There is no record of a Wintul Berry in this regiment, but it could be a nom de plume for William E. Barry who enlisted in the unit later named Company G on July 19 1861 in Harrisburg, Texas at the age of 22. He was wounded in one arm at the Battle of Gaines Mill and captured at the Battle of Antietam. After being exchanged from the POW camp, Fort Delaware, Barry was elected 3rd lieutenant in March, 1864. He was wounded in the arm at the Battle of the Wilderness on May 6, 1864 and resigned his commission and retired due to disability on November 28, 1864.

Private William L. "W. L." Edwards
4th Texas Infantry

Richmond Va. May 10, 1862

I will write you a few lines again which leaves me in bad health, I had a light fever yesterday and feel very poorly today, but I hope I will be well in a few days. There is over a thousand men in this camp and out of that number, I do not think there is a hundred that have not a bad cough, and it is cough, cough all night long. I tell you there is little fun or pleasure to be seen here. The hospital is crowded with sick soldiers ever since I reached here but have not been very sick. I have been sick with the flux for the last 4 or 5 days but think I am getting well now. The soldiers life is a tolerable hard one, but I will not complain; I have had plenty to eat, and hope that I will still continue to have. I would love to be where you are and [see] what you are doing. I think it is time I should get a letter from you. I want you to write whether you have heard from Uncle Ashley or Onslow yet, and give me all the news you can that would be interesting to me. I have no news to write you, we soldiers are kept in the dark as much as possible. We have not had any battle since I came here, and I have no idea when we will.

I am probably as well satisfied here as I would be if I was there, under existing circumstances for I do not think I could be satisfied now, anywhere. We have had as good a Captain as I would have. I am pleased with him, as well as the rest of the company; they all seem to be brotherly and kind to each other; though nearly all of them are tolerable wicked men. Roxie, I want you to pray continually for me, and for our cause, and that this war may speedily come to a close; and I believe that it is the duty of every Christian to do the same, "for fervent affectual prayer of the righteous avaleth much." We can do nothing if we rely upon our own strength; for our help must come from God! And unless we humble ourselves and come to Him with our supplications and prayer we can do nothing. Oh, how I would love to be with you and my little boy this morning; I still hope and pray that this war will soon close, and that we will be permitted to return home in peace. "We may be happy yet my love, we may be happy yet."

I have written nearly all that I know to write. I have three dirty shirts and one payer of drawers that [I] will have to go and wash. I bought one hickory shirt which cost me 75 cents. I have plenty of clothes to last me six months if I do not have to throw them away; but if we have to make a force march we cannot carry our knapsacks. I have seen good numbers of good and they run from 10 to 15 a day. Old Uncle Bill Carter has been sick ever since we left N. Orleans. We have orders to march tomorrow toward the yankees; Our army was engaged yesterday and we lost 1600 men, and killed 6000 of the yankees and took 800 prisoners. We have drawn our guns, knapsacks, & etc., we do not know what day or hour we will be called on to fight; for the yankees are marching on toward this place with the determination of taking it, but I trust to my Creator that they will make a failure, if they get this city I fear we [are] ruined; I hope and pray that this war will soon end for I do not feel like I could stand it one year much less three. I do not know that I will be able to leave here tomorrow, but I will go if I possibly can for I would much rather be killed on the battle field than die in the Hospital. I do not know when I will have an opportunity of writing to you again. It may be possible that this is the last lett4er you will get from me; though I will write whenever I have a chance to mail my letter. I want you to keep writing. Until I can get a letter from you, for nothing would give me more satisfaction now. If I can get home alive I will feel satisfied; the boys that have been here nearly a year say it takes all the money they make to get things that they are obliged to have; but I will try and save a few dollars to bring home with me if I am so lucky to ever get back. I want you to live in the faithful discharge of your duty, live prayerful and watchful, and try to raise our little boy right. I feel I like I will be spared to see you again. I can not send you stamps for it is so uncertain about you getting them. I will write you as often as I can. For the present I will close.

Your affectionate husband

W. L. Edwards
Direct your letters: W. L. Edwards Co. "K"
4th Texas Regiment
Care of Capt. Martin
Richmond Va[12]

---

12. Letter, William L. "W. L." Edwards to Roxie Edwards, May 13, 1862. Hood Texas Brigade files. Texas Heritage Museum-Historical Research Center.

Private W. L. Edwards
4th Texas Infantry

To Travis and Elizabeth Scott

Dear Uncle and Aunt

I must write you a few lines. I have the worst cough that I have had in several years and nearly everyone in our company has the cough. It is said that the smallpox is in camp but I have not seen any person that has it. I hope you are all well and will make plenty of corn. There has been a great deal of rain in this country and most of the corn is just coming up. I tell you it is hard times here. It is believe that this place [Richmond] will be taken in less than two weeks, but it is the least of my opinion. Our boys, Martin's Co., have been [fighting] this week, but I have not heard whether any of them are killed, but I guess several of them are killed. I expect there will be [fighting] this week, but I have not heard whether any of them are killed, but I guess several of them are killed. I expect there will be [fighting] until the matter is decided.

Your Nephew

W. L. Edwards[13]

Private W. L. Edwards
4th Texas Infantry

In Camps Va. May 13, 1862,

Dear Roxie,

It has been but a few days since I wrote to you and consequently I have but little news to write. We left Richmond Sunday the 11th and are now in camps about 20 miles east of Richmond; our boys were engaged in a battle a few days ago and killed several of the enemy but none of the companies were killed and but one wounded. I wrote in my last letter to you that our boys had been in a battle and that we had killed about 16 hundred of the enemy but this is a false report, I guess that there was over two hundred of the enemy killed and some 30

13. Letter, W. L. Edwards to Travis and Elizabeth Scott, May 10, 1862. Hood's Texas Brigade Files, Texas Heritage Museum-Historical Research Center. The letter was written on the back of a letter on May 10, 1862 to Roxie Edwards.

or 40 of our men killed. When we reached here Sunday evening the Yankees were a firing from their gunboats. We have been here nearly two days waiting for and expecting an attack; we are expecting a fight every hour the enemy is within 3 miles of us and has been for 4 days. I have just heard that the Yankees are advancing on us there is no telling.

My later dated the 20th as you see and this is the 24th. I feel tolerable well this morning. I hope you and little Joe F. are well. While I am writing this, the canons are roaring but that is not anything new here. I want to get a letter from you the worst you ever saw. Goodbye my love your husband until death.

W. L. Edwards

Roxie if I get killed in battle I do not want you ever to marry.[14]

Private W. L. Edwards
4th Texas Infantry

June 2, 1862

My affectionate companion:

I have not received any letter from you nor heard from you since I left, and you can guess how anxious I am to hear from you. I am well this evening but I have been sick some since I came here; I have no news of importance to write to you. One of our men has died since we came here. Uncle Billy Cater can tell you all the news about the war. We have had a hard time of it now for the last three days, for we have been moving about almost all the time for three days, and we expect a great battle tomorrow.

I ask that my God will preserve me from danger and harm, and if I am killed I hope that I will be taken to a land of rest where war, sorrow, and trouble never come. I feel that I can put my trust in God, and if he will be pleased to spare me, I hope that we will get back by Christmas if not sooner; I cannot think that the war will last more than a month longer. I have some trophies which I got on the battlefield. I have a fine bowie knife, a fine coat for which I have been offered twenty dollars and several other valuable articles. This sheet of

14. Letter, W. L. Edwards to Roxie Edwards, May 13, 1862, Hood's Texas Brigade files, Texas Heritage Museum-Historical Research Center. Private Edwards was mortally wounded in the battle of Gaines Mill fought on June 27, 1862. He died in a military hospital near Richmond, Virginia on July 18, 1862.

paper was in a portfolio which I took, and a great deal more besides. I also will send you an envelope. I hope we we'll meet again, but if not I feel resigned to the will of God. My greatest reason for desiring to live is on yours and our little Joe's account; if you were both gone I feel like I could willingly go. We are called in now to fall into ranks and I will leave to quit writing. Roxie my darling wife I want you to live humble, prayerful, and never forget your duty toward your God and train up our little boy in the way he should go. Oh, Roxie if the good Lord spares me to get back I shall lead a more devoted life to him [than] I ever have before pray earnestly that we may live to meet again on earth but if we do not, oh; may we meet in heaven.

Roxie I wrote to you that I did not want you to marry if I got killed but perhaps that is a foolish request; use your pleasures about that; but let's not talk about that, I would try and send you some money but I think it would be unsafe for it is uncertain whether Uncle Billy gets back or not.

For the present Goodbye.

From your affectionate husband

W. L. Edwards.[15]

Private Mark S. Womack
4th Texas Infantry

On the morning of the engagement, we had hardly proceeded a mile from where we had camped the night before, and were marching along carelessly with empty guns, not suspecting a foe. The first indication we had of the enemy being near us, was a rapid firing at the head of the regiment and the whizzing of balls around us. One of Captain Porter's men was wounded at this fire. In an instant all was excitement. The order was given to load and form into line of battle, which was promptly done. The enemy after firing upon, retreated across an old field back into the woods. Five of them were shot down by some of Capt. Key's company at a distance which could not possibly have been less than six hundred yards. The regiment being quickly formed into line of battle, then it was that Captain Carter's company, and ours were sent forward into the thick woods as skirmishers as mentioned in Capt. Hutcheson's report. Ours was a

15. Letter, W. L. Edwards to Roxie Edwards, June 2, 1862, Hood's Texas Brigade Fills, Texas Heritage Museum-Historical Research Center

brilliant little skirmish - the company did more execution that day than any other single company upon the field, from the fact that it was in a better position to do it than any other. Capt. H. put down the loss of the enemy in killed and wounded at the lowest figure. Maj. Warwick, who was with us all the time and had as good an opportunity of judging, says that their loss, killed and wounded cannot be ascertained, for they were lying through the woods in every dircction. But what was most remarkable, we lost not a man, only one wounded, thigh broken.

Private Mark S. Womack, 4th Texas Infantry.

(Texas Heritage Museum Historical Research Center)

The boys all behaved nobly under the enemy's fire, never flinching, but moving forward amidst the whistling of balls shooting down the foe with deadly aim until they were completely routed and driven from the woods. What the force of the enemy was, I cannot say; but presume it was a full brigade that encountered ours. Other brigades were near to support us in case of necessity, but we needed no assistance - we met the foe in equal numbers and upon equal footing and drove him back. The 5th Texas were on our right and had a skirmish with them. The 1st Texas were on our left inline of battle and received the fire and charge of a whole regiment - they wavered not, but poured a deadly volley into the enemy and charged them in return and drove them back towards their gunboats. Being then under secure shelter, they poured bombshells and grape into the woods where we were - they whizzed and fell in close proximity, but nobody got hurt. We then retired from the field, perfectly content with our day's work - the boys bearing away many trophies of the fight.

Texas had a fair showing on that day - no other regiments were engaged but the three Texas regiments. It was emphatically a Texas fight. We fought our own way, and upon our own ground, (except that the enemy had little choice of position). It was a Bush-wacking fight, every man fought as he pleased and we won the field.

We are upon the threshold of stirring events. The main conflict must take place before long. Living upon (?) rations a day – the boys in fine spirits – the army (?) shall all hear from us again.

Mark S. Womack[16]

Private Bennett Wood
4th Texas Infantry

After the battle of Williamsburg the brigade pushed on to Eltham's Landing, where the enemy was landing a force from gunboats and transports, evidently with the object of capturing the rear of the column with the baggage train. The Texans and their advancing line, in a frolicsome way, soon had them under their boots; quite a number were killed and wounded and captured, between 400 and 500. After a wet muddy tramp we took a rest below Richmond. Early in June we were on the move, it was said, to join Jackson in his strenuous operations in the Shenandoah Valley.[17]

---

16. Private Womack's letter about the battle of Eltham's Landing was found in the scrapbook of his niece, F. B. Womack, which she made during the Civil War. Mark S. Womack to Unknown, circa May 1862, Hood's Texas Brigade files, Texas Heritage Museum-Historical Research Center. Mark S. Womack He is listed as being sick in 1862, and was captured at the battle of Gettysburg July 2, 1863. He was sent as a prisoner of war to Fort Delaware and was paroled on June 9, 1865.
17. Yeary, boys in gray, 815. He was wounded at the battle of Gaines' Mill on June 27, 1862 and also at the battle of Gettysburg on July 2, 1863. Wood was promoted to 3rd Sergeant in March 1864 and was wounded in the foot at the battle of the Wilderness on May 6, 1864.

Anonymous
4th Texas Infantry

Galveston Daily News
December 6, 1896

The Bride of York River
A True War Story
by Luther Coyner, San Diego, TX.

While the writer was enjoying a smoke on the porch with a well-to-do farmer, who lives not more than 200 miles from Galveston, several years ago, the worthy host remarked:

"So you have heard of "The Bride of York River? Yes, she is in Texas. And you would be pleased to see her and hear the true story which caused so many newspaper comments in 1862-1863? Well, I will relate it to you."

Thus followed the story, which the writer gives in as near the farmer's words as can be remembered:

"The Bride of York River was born near West Point, on York river, in King William County, Virginia, May 4, 1844, of wealthy parents. Her grandfather had built the elegant brick mansion where she, Millie, was born, soon after the revolutionary war.

He had been an officer in the war of 1812 also, and Millie's great grandfather was a student at Princeton College and was one of the first to enlist in the war, and was a private under Washington, and was at the siege and severely wounded at the battle of Yorktown.

Millie, the bride of York River, was an only child and lived with her mother and servants at the old homestead in 1862.

At this time, when my story opens, her father was commanding a regiment in the Confederate army, having raised and commanded a company in 1861.

My father had moved from Culpeper County, Virginia, to Texas when I was small, and I was in my 20th year in 1861 when I joined the Fourth Texas Infantry regiment, and in May 1862, was a private in Carter's company of that regiment, then commanded by Colonel John Marshall, who gave up his noble life at Gaines' Farm.

The other two Texas regiments in the Virginia army – the First and Fifth, were commanded by Colonels Rainey and Archer, afterward,

General Archer, and belonged to General Hood's brigade, General Whiting's division.

On May 7, early in the morning, the brigade was ordered to march on the road leading from New Kent C. H. to Eltham's landing, on the Pamunkey River, with the exception of the Fifth Texas regiment, which was ordered to take the Blind road leading to that landing.

On arriving within a short distance of our cavalry pickets the enemy suddenly appeared deployed as skirmishers and immediately opened fire on us. We were in front and the guns of the whole Fourth regiment were unloaded, as we had not been ordered to load and were inside the line of our pickets. We were ordered to load and did so under fire. Then came the order that the Fourth Texas move forward as skirmishers, and as we moved forward as skirmishers, and as we moved forward, driving the enemy through a dense forest, was when I first saw the bride of York River.

I had entered the woods and advanced some distance when in a by-path within twenty paces ahead of me I saw a sight that not only surprised and astonished me, but drew forth all my rebel instincts and led me to do what afterward caused me worry, but finally resulted in much pleasure and happiness.

In the path in front, where the two had met, were a lady and a Federal officer, both too much engaged to notice me had I not been hid by the trees that intervened.

The lady was dressed in a near fitting blue riding habit and jaunty hat and was mounted on a fine Morgan filly that seemed to be under her control. The form of the young lady was perfect and the side of her face, I only could see, showed the perfect Southern beauty of which I had read so much about but before never had seen.

In fact, for a few moments I never saw, heard anything or did anything but gaze at this wild flower of the Virginia woods.

The officer was also mounted and from the first words I heard him utter they had met before, and I afterward learned it was the young lady's home, where General Franklin, the commanding general, had his headquarters. The officer said:

"Aha! Miss Millie, I have caught you at it. I learned last night you were bent on mischief and have been sent to find and capture you and take you before General Franklin."

They were then facing each other in the path, which was barely wide enough for two to pass.

"Let me pass, sir!" said the young lady, "I don't believe a word you say about General Franklin: surely he's too much of a gentleman to make war on defenseless women, even if he has taken their home for his quarters, and that, too without their consent."

With a bitter laugh the officer said:

"But he has learned that you intend to betray him and has ordered your arrest: but (with a somewhat softer tone), Miss Millie, if you will go with me and be my southern bride, as you promised three years ago, he will never know that I have found you and all will be well."

The young lady raised her riding whip and said with scorn:

"Traitor! Base traitor! You would even betray your general!: let me pass; I defy you!"

Then, suiting the action to the word, the officer, as he said:

"Then I will take you by force." And spurring his horse alongside that of the young lady he dodge the whip (she having struck at him) and grabbing the young lady around the waist drew her from her pony, which became frightened and galloped away.

All this had occurred in less time than it takes me to tell it, I had raised my rifle and at once cried "Halt!"

Turning his face toward me for a moment the officer immediately released his hold on the young lady, who cried out:

"Oh! Save me, save me!"

At the same time she bent low enough to be out of range of any shot. I had noticed the quick movement of the officer to spur his horse and on the instant I fired.

Both the officer and the young lady dropped to the ground and the other horse galloped away.

When I reached the spot where the dead officer lay, for I had aimed well, the young lady had risen, and was standing over her recent foe, pale but defiant.

As I came up she noticed my uniform was gray, and said:

"Oh, thank you sir, thank you!"

I realized the situation she was in, and urged her to move backward toward the rear of our troops, and to ask for General Hood.

With the assurance that I would explain matters, I gave her my name to give to General Hood and she bravely started for the rear.

By all this I had lost my place in line, but our regiment, being supported by the First Texas, we drove the enemy a considerable distance, when they were reinforced, and we were attacked with such a terrible fire on our flank by two or three regiments of the enemy.

Our regiment was then ordered to charge and, after driving them about a mile through the most difficult forest I ever say, forced him under the protection of his gunboats, and about 2 o'clock p.m. we were ordered to retire.

I could give you much more of this day's battle, but that is not the object of this story.

History tells of the fine work the noble Texans did on that day, and on many other bloody battles of this cruel war.

Need I tell you that I fought harder than ever before, and need I say that there was ever a sweet face before me?

At our camp when I returned, I found an order awaiting me from General Hood to report to him at his headquarters.

At the general's tent I found Miss Mille and nearby her pony was hitched, which, with the Federal officer's horse, had been captured by our troops.

Miss Millie had evidently told the general all, for General Hood called me by name and said:

"I know all the circumstances of the rescue of this young lady, and have the important papers she brought, and to reward you for your bravery and forethought, as I need another orderly, you shall have the position.

This, I assure you, was a "ten-strike" for me, and I noticed that Miss Millie seemed more pleased than I did, and I think the general noticed that fact too: anyway, he took occasion to leave us there alone sometime.

As soon as we were alone, I brought up the subject of my firing upon the Federal officer, for I was afraid that she might think I had been too hasty, and that the fact of me killing him was not exactly under the rules of civilized war.

Mille spoke frankly and said that she thought I had done exactly right: that this man had not only been a traitor to his own state, but had harassed and slandered her for the last two years and had circulated the report that she had secretly married him, calling her his "bride of York river," and had lied to General Franklin, telling him that her father's house belonged to him, and while it was true that before the war he had visited at her father's house, she never had promised to marry him, for she was then a mere child and after she had learned he had joined the Federal army, and within the last few months learned of the scandals he had circulated, she had ceased to even suspect him and almost hated him. "Besides," she continued, "he was in our lines, and was trying to capture a loyal subject, and had you not fired when you did and your aim been at fault, he would have succeeded in not only carrying me off, but the important papers I had for General Hood."

When the general returned he promised to protect as far as he could the young lady and her mother, and by the aid of a servant, Mille and her mother were able to reach Richmond in a few days, from which place I frequently received letters from her until the battle of Gaines' Mill.

When that bloody battle occurred I thought my time had scarcely come. I was severely wounded, and was finally taken to Richmond. I had not been there long before Miss Mille learned of the fact, and when I was able, or thought I was able, I was taken to her mother's private house. The move proved too much for me, and for weeks I lingered between life and death, and it was months before I was sufficiently recovered to join the army in the field.

One day in the spring of 1863 Miss Millie and I were waling among the graves of Hollywood, when Miss Mille grasped my arm and said:

"Look! Who is that?"

Her face was deadly pale, and looking in the direction she pointed, I understood the cause. Not ten paces away stood the man who had claimed her as his bride of York River, and the man I thought I had sent to his long home nearly a year ago.

He was dressed in civilian clothes, much paler, with eyes sunken and cheeks hollow. Sure enough, he stood with arms folded and eyes shining, gazing upon us with a look of scorn and hatred, that in the fading sunset of that spring evening fairly seemed to make my blood run cold. Naturally I grasped my own sword, for I had donned my uniform, and this I expected to report for duty and on the morrow leave my promised bride for the field of duty.

Mille clung to my arm, when I said:

"I am glad he is not dead; I will speak to him."

We advanced, Millie holding my arm, nearer to where the stranger stood. He never changed his attitude, but stood looking at us with the same look of scorn and hatred, when I addressed him and said:

"Good evening, stranger; then you are not dead?"

"No, but you are no less a murderer at heart. It is not your fault I am alive today, Guard well your promised bride of York River."

He then turned on his heel and was soon out of sight. For several minutes Millie and I stood there in silence, and then, without saying a word, started toward the city.

I felt her hand trembling on my arm, and I tried to talk of other things, but she brought the subject back to the recent occurrence and begged me not to go back to the army and said that the stranger would surely follow me and do me some harm

I found out the next day, before I left Richmond, that he was still a prisoner of war and that he wound was of such a nature that he was unfitted for action in the field, which to some extent pacified Miss Milled.

I returned the next day to my command. Then followed the eventful year of 1863. "On to Gettysburg," and the return and the battles of 1864 and finally Appomattox. It is not necessary to go through all the details, though I could tell enough to fill a book. My letters to and sent from Mille got scarcer and less frequent. Not on account of the feeling between us changing, but because we could not get paper to write on, and the army in the spring of 1865 was constantly on the move.

Neither of us had seen or heard of the wounded Federal officer since that evening in Hollywood cemetery.

On April 15, 1865, poor, ragged and down-hearted, I started for Richmond, without any object except to see Millie and talk over future prospects.

Reaching the once glorious, but now dilapidated city. I soon found the home of Millie. She met me as she always had done, in her happy, cheerful manner, and said:

"Oh! Now, you can stay with me always?"

She then told me of her sorrow. Her father had laid down his noble life at Trivillion and her mother was now a confirmed invalid, caused by an accident in the street with a runaway team. "But now," she said, "I have you with me, safe and sound. No more war, no more fighting, but peace, peace!"

We decided to have a quiet wedding and move to the old place on York river. So on May 4 (Millie's birthday), 1865, the dear little minister, who had stood so nobly by the Confederacy in its dark days, and has now preached over fifty years for that people, made Millie and your humble servant me and wife.

Millie look lovely in her home made gown, and the groom had on a homemade pair of pants, but a "soiled shirt," if you please, if he did not have on any coat.

On the next day we went out for and on the 12th reached West Point, for we traveled by wagon, or rather by a one-horse cart, and on the 15th we were somehow established in the brick house, which was about all that was left of the estate. Fences and other improvements were all destroyed.

Some of the negroes were still in on the place, and wanted to remain. So we had what help we needed.

The trip of Millie's mother from Richmond was too much for her shattered frame, and on the last day of June of this year we laid the dear old lady to rest.

This somewhat dissatisfied Millie, and she suggested that we do what I asked her to do in May, go to Texas.

While Millie and I were sitting on a front of the old brick house on the evening of the 15th of August of that year we were talking about going to Texas when a poor, sick stranger hobbling toward my house, Millie said:

"Some poor, crippled Confederate soldier came by and I asked him in to get something to eat."

She went out and as the man approached to my surprise, I noticed that our guest was no less than the Federal officer who had given Mille the name of "The Bride of York River."

I asked him in and set a chair and told him Millie had gone to get him something to eat. He came in and thanked me and sat down but said he was too far gone to eat.

Just then Millie came, but came near dropping the tray when she saw who was there. She soon gained courage and offered the sick man something to eat. He refused saying he needed nothing in this world and had not long to stay, but that he had heard we were married and before he did he wanted to bless the union, and do what he could to remedy any wrong he had done.

"Now I will go," he said, "since I have accomplished my object."

He attempted to rise, but both Millie and I insisted upon his staying with us, but he said "No, I must go."

The exertion and excitement was too much for his feeble frame, and he fainted away, and would have fallen had I not caught him.

I assisted him to a room and laid him on a bed and did what could be done for him. His time had come and on the second day he passed away never having gained consciousness after he had fainted on the porch."

This is the story.

"We sold out in the fall of 1866 and moved to Texas, arriving at Galveston September of that year.

We bought this place in 1867 and have been in grand old Texas ever since.

Now you have the story of "The Bride of York River."

Here she comes now to invite us to dinner, let us drop the subject.

Though the bride had seen over fifty summers come and go, she was still beautiful. She has a married daughter residing in Houston. Another daughter is attending school in Virginia, and two sons, one in Texas, a lawyer, and the other is reading medicine in Louisville, Ky.

The husband of "The Bride of York River" is a factor in his community, proud of his four years under Lee, proud of his grand old Texas, proud of his brave, noble and beautiful wife, proud of his children, and proud of the action that gave him "The Bride of York River."

Anonymous
4th Texas Infantry

Western Sentinel
June 6, 1862

We have been kindly permitted to publish the following extracts from a private letter, written to his relatives, by a soldier in the Texas Brigade. He is a native of Salem, N.C., and his gallant conduct so modestly related, cannot we think, prove uninteresting to our readers:

Camp Texas Regiment
Near Richmond, Va.,
May, 25, 1862.

Our division of the army assisted in covering the retreat from Yorktown, and our Texas Brigade had quite a considerable little fight with the Yankees near a little town called Barhamsville, in New Kent County. The enemy landed in force at West Point on the York river, and endeavored to cut us off from the main body of the army. We were ordered to engage them, and for this purpose left the main road and marched in towards the river. When within about 2 miles of the river we were fired upon by the pickets of the Yankees and one of our men slightly wounded. Our regiment the 4th Texas, immediately ran forward into line of battle, our first Company firing upon the pickets as they ran off, killing four and taking one prisoner. This occurred in an old field, to our right and in our front were thick woods. In a few minutes our Company and another were ordered by our General to march into the woods and deploy as skirmishers. This we did and it was but a short time before we came upon the Yankees, and went to fighting in true earnest. We took trees, stumps, &c., on them, and soon set them to running. We were thus employed skirmishing for about three hours. In the meantime three Regiment of Yankees had by a flank movement in our rear and we were about to be cut off, but our General was equal to the emergency. The first Texas Regt. met them and after firing three vollies, charged them at the point of the bayonet and completely routed them, driving them back to their gunboats. This ended the fight. During the engagement they shelled us from their boats, but did not succeed in doing any damage. We got all of our dead and wounded off the field before night and then before day continued on our march. Yankee reports of the engagement say it was the hottest fight on the Peninsula. They say that 20,000 of their troops were engaged and they acknowledge a loss of 300 in killed and wounded. Our loss was but little. Our three Texas Regiments were the only three engaged, the 1st lost in killed and wounded 16, the 5th had 2 killed and our Regiment had but 2 wounded.

Our Company behaved most gallantly, we killed about 30 and took 22 prisoners. I, myself, killed 1 dead, broke another's leg and took 2 prisoners. I felt sorry for the poor fellow whose leg I broke, so much so, that I tied it up for him, gave him a drink of water and put him in as comfortable a position as I conveniently could. This was my first fight and if it were possible, I would like for it to be the last. But enough of this.

I can assure you that we saw rather hard times in our recent marches. We have best nearly starved to death. We would get for our days rations about one third a pound of fat bacon, and two crackers and a half to the man. Sometimes we would get a little flour, and as we had no cooking utensils, we had to take the bark of a tree to make up our dough in, and then wrap it around a stick and hold it over the fire to bake. We made some very good bread in this way. But we are doing well now. We are getting plenty to eat and occasionally draw such vegetables as collards, potatoes, onions, &c. We have a wagon detailed to go to Richmond to the market every morning and we can purchase almost anything there

---

Anonymous
4th Texas Infantry

"We marched out of camp that morning at daylight, each of us wondering where we were going, and not a soul of us suspecting that an enemy was near. We went about a mile and a half, and the Fourth Texas in advance, were passing through a field dotted with pine stumps, and approaching a house situated on the crest of the hill overlooking York River valley. Hood and some members of his staff, and perhaps a courier or two, rode about fifty yards ahead of Company A, the leading company. To the left of the road, and about forty yards from the house, sat a cavalryman, apparently fast asleep on the back of his stead. Hood rode on by him, but twenty in number, sprang from behind the house, and fired a volley at us. For a second, consternation prevailed. Not a man of us had his gun loaded, and there was a pell-mell scattering to take shelter behind the many stumps. Hood wheeled his horse, and shouting, 'Fall into line men – fall into line,' came dashing at full speed toward us. Halfway to us, he noticed that nobody was paying any attention to him, and he shouted 'Get' into line, men – get into line, Fourth Texas! Is my old regiment going to play hell right here?' Just then a Yankee stepped out in front behind the house, and seeing him, John Deal – the one and only one of us that had a loaded gun –dropped to his knees, took careful aim and laid the daring fellow low. In another second every gun in the command was loaded, and the men began moving into line, and having formed a semblance of one, rushed forward to take the crest of the hill and commenced firing at the Yankees now in swift retreat across an open field in the valley between us and heavily timbered land. The other regiments of our brigade moved quickly up on the right of the Fourth, and within five minutes such of the Federal skirmishers as were not killed or wounded had fallen back to the protection of the timber.

"Hood then ordered forward a skirmish line, and following it, we crossed the open field in the valley and gained the timber beyond. Then, the Fourth Texas, the only regiment whose movements I know anything about, began hunting for the enemy. But although there was an abundance of Yankees near us, as was evident from the firing on our right and left, not one of them appeared in front of the Fourth. Its only loss was from the volley fired at us on top of the hill. By that, one man was seriously wounded. The same volley, I have always understood, killed Captain Denny, the commissary of the Fifth Teas. As to that, I cannot speak positively; I was having my first experience in being shot at, and was therefore observant of only what occurred in my own regiment, and near at hand. In saying that the other regiments of the brigade moved quickly up on the right of the Fourth Texas, I may be in error, for one or the other of them might have formed of on its left."

Eltham's Landing – by Chuck Haas.

General Hood saw, before the advance, that there was danger of confused firing in the woods and he was determined not to permit his men to lad their guns until they reached the Confederate cavalry picket. He then rode ahead of his column, which had been moving forward by the left flank. His own beloved 4th Texas was in front. When Hood reached the cavalry picket in the rear of a little cabin, he ran squarely into a heavy Federal skirmish line. Behind the skirmishers was a heavy force. Instantly Hood jumped from his horse and ran back to his own troops, who fortunately were not fifty feet away. One Union corporal picked out the figure of Hood in front of the regiment and deliberately drew down his rifle on the general. A military career of high promise seemed at its end - because the muskets of the disciplined 4th Texas obediently were unloaded. At that instant a single shot was fired, but it was the Federal corporal, not Hood, who was killed. One strong-minded individualist of the regiment, distrusting orders to approach with an empty gun, had loaded his piece before starting and by instant aim, saved Hood's life. This soldier was John Deal of Company A, who survived the war and lived subsequently at Gonzales.[18]

Raleigh Register
May 31, 1862

From the Richmond Enquirer.

West Point.

Inscribed To General Hood.

By Chas. S. Worsham.

---

18. Chuck Haas, Hood's Texans – "An irresistible Attacking Force," in *The Junior Historian*, December 1965, Vol.26, 3.

Earth was hush'd in quiet slumber,
Stars were beaming from the sky,
Soldiers all lay sweetly dreaming,
Though the foe were lurking nigh,
Fires around were dimly burning,
Marking where the sleepers lay,
Clad in armor waiting ready,
Should the call them to the fray.
Winds were sighing soft and lowly
Through the branches of the trees,
Lulling with their tender music
Wearied souls of Southern braves.

Hark! For through the twilight stillness
Comes the cannon's sullen roar,
Rousing many a gallant hero
With its thund'ring voice of war.
Quick, to arms! the word came passing
Rapidly along the line;
Forward, forward! foes are coming,
We, too, must be there in time.
Hurrying onward, soon our columns
Reach'd the densely studded wood,
Where, drawn up in line of battle,
Lincoln's bands of hirelings stood.

Charge them! cried our noble leader,
And we rush'd against them there,
Scatt'ring them like leaves of autumn,
Through the swamps and everywhere
Musketry now crash'd and echoed
Through the dim aisles of the wood,
And where fiercest rag'd the conflict,
There was seen the gallant Hood

But their last ranks are now flying,
Dead and dying strew the ground;
Walling cries go up to Heaven,
From the mangled suff'rers round.

Again night's mantle dark enshrouded
Earth within her gloomy fold;
And the stars again beam'd brightly
On dead foes, pale and cold.
Many a proud and youthful bosom,
That once glow'd with patriot fire;
Now lay dead, their faces gleaming,
Through the dusky twilight air.
Hurrah! hurrah! the lone star banner
Waves, though in the night's dark shade;
Hurrah! for hood, our gallant leader,
And the Texas Light Brigade.
Co. E., 4th Texas Reg't, near Richmond, 1862.

5th Texas Infantry flown at Eltham's Landing

(State of Texas Archives)

CHAPTER THREE

5th Texas Infantry Regiment

The Regiment was authorized by the War Department of the Confederacy on October 1, 1861.

List of Companies and First Company Commanders

Company A: (Bayou City Guards) Harris County - W. B. Botts
Company B: Colorado County - J. C. Upton
Company C: (Leon Hunters) Leon County - D. M. Whaley
Company D: (Waverly Confederates) Walker County - R. M. Powell
Company E: (Dixie Blues) Washington County - J. D. Rogers
Company F: (Invincibles No. 1) Washington County - King Bryan
Company G: (Milam County Greys) Milam County - J. C. Rogers
Company H: (Texas Polk Rifles) Polk County - J. S. Cleveland
Company I: (Texas Aids) Washington County - J. B. Robertson
Company K: (Polk County Flying Artillery) Polk County
I. N. M. "Ike" Turner

The 5th Texas Regiment left Texas in 1861 with approximately 800 enlisted men and officers. During the winter of 1861-62, before the regiment's first engagement, the ranks of the 5th Texas shrank significantly as illnesses swept through the winter camps.

In order to bolster their strength, in early spring 1862, an officer and enlisted man from every company was sent back to Texas to recruit in their home counties. The large size of the 5th Texas during the first two major battles of the 1862 campaign reflected the success of this recruiting effort.

Brigadier James J. Archer, as captain of infantry during the Mexican War, received a promotion to brevet major for gallantry at Chapultepec. He was honorably discharged on August 31, 1848. Resigning his commission in 1861, he was made a colonel commanding

Colonel James J. Archer, Commanding officer of the 5th Texas Infantry.

(Library of Congress)

the 5th Texas Infantry. He was promoted to brigadier general, and transferred to lead the Tennessee brigade. Though promoted and transferred Archer and the men of the 5th Texas lamented his transfer and he always remembered his command of the 5th Texas with pride and admiration. On July 1, 1863, at the battle of Gettysburg, Archer and a large part of his command were captured. He became a prisoner of war for more than a year before he was exchanged in the summer of 1864, with his health being shattered by his confinement on Johnson's Island. Posted for duty with the Army of Tennessee, on August 9th, and ten days later with the Army of Northern Virginia, he commanded for a short time his old brigade and that of Gen. H. H. Walker. He died on October 24, 1864, in Richmond and is buried there in Hollywood Cemetery.

Adjutant Campbell Wood
5th Texas Infantry

Daily Express
November 24, 1907

Much to our delight we remained in the vicinity of Rocketts but a few days being moved to Camp Bragg, near Braggs Mill, about three miles from Richmond. Here after all the companies of the regiment arrived, the field officers appointed to command us by the Confederate authorities came to us. For Colonel we were given J. J. Archer of Maryland; for Lieutenant Colonel, Lewis Addison Armistead of Virginia, and for major Quattlebaum – all three splendid officers, honorable gentlemen and gallant soldiers who held rank in the old army. Each well equipped to command us and make good soldiers of us, they were the very men we needed. But notwithstanding that our men knew the terms which we had been allowed to come to Virginia, they opposed these appointments strenuously and clamored for Texians as field officers. Learning the opposition to him, Colonel Armistead immediately went to Richmond, and in person applied for transfer to another command and much to our sorrow in later days, his request was granted. Like who apprised of the discontent prevailing, Major Quattlebaum forwarded his application for a transfer from camp. Action was delayed, but finally his request also was granted.

Adjutant Campbell Wood, 5th Texas Infantry.

(Elisa Vorwook-Wood)

Colonel Archer was less sensitive than his subordinates and far more sensible. Knowing his own merits and trusting to them to win him the confidence of the Texians, he stayed with the regiment. His faith in himself was amply justified. On the morning of our first battle, that of Eltham's Landing, when he rode out in front of the regiment and saying "Fifth Texas, I have sought and obtained permission for you to open the ball this morning." Immediately, gave the command, "Forward!" our fellows began to realize that their dignified Colonel, although short of stature and small of body, had in his heart the spirit of the true soldier. When during the battle that followed, we thought he was exposing himself unnecessarily, one of the men approached him and said, "Colonel, we can't afford to lose you now, get behind a tree, sir, we will do the work. If one of us is killed there is another to take his place, but there is nobody that can take yours." Such solicitude in his behalf on the part of the men that so lately had not a good word to speak in his favor, affected Colonel Archer deeply, and he spoke of it often in later days. When during the battle of Seven Pines he was made a Brigadier General and ordered to another command at once, he again faced the regiment, as it stood half-leg deep in water, and said: "Fifth

Texas, I have come to bid you good-bye this morning," and would have continued had not his emotions overcome him, and he galloped off without another word. Not a dry eye was in the regiment, for it had come to love and respect the gallant little man and to have the utmost confidence in him.

Nor was it affection unreciprocated. After every other battle in which the Fifth was engaged, General Archer either came in person, or sent a member of his staff to inquire how the regiment had fared.

Adjutant Campbell Wood as elected third lieutenant of Company D. In June 1862 he was appointed to the position of regimental adjutant and was wounded at Gettysburg, losing three toes. He survived the war and became a physician in San Saba County, Texas.

Adjutant Campbell Wood
5th Texas Infantry

I am positively certain that the Fifth Texas was assigned to the duty of opening the battle. I am equally certain that no other regiment went in with us, or was, at any time during the battle aligned with us. The Fifth was drawn up in line in an old field or meadow back of a little village, when, riding out in front of it, Colonel Archer said: 'Fifth Texas, I have sought and obtained permission for you to open the ball this morning.' Then he gave the orders, 'Right face, file left, - Forward! March!' We moved across the opening, and soon struck the timber, taking a road on each side of which was a dense thicket of undergrowth in full leaf. No skirmishers or scouts advanced in front of us, and judging from that fact, I would not believe a battle was imminent. I forgot to say that before giving the command, "Forward,' Colonel Archer ordered the men to load their guns, but not to cap them.

"We marched down the road toward West Point in column of four ranks, Colonel Archer and Captain Denny, rode at the head of the regiment, and I trudged along on foot, immediately behind Archer and Denny rode slowly, and the men kept close up, talking a little as was usual on a march. Just as we approached quite near to an old shack of a house, a Federal sergeant and eight men jumped from behind it and fired a volley at us. All their bullets excepting one went wild, but that one struck and killed Captain Denny, and he fell from his horse. Archer immediately ordered the men to cap their guns. Many of them, however, had capped theirs when they loaded them, and these men sprang from the ranks and fired at our assailants, killing and wounding every one of them.

"Breaking into a double-quick, and following close on the heels of two or three men Archer had ordered to keep well in advance of us, we went forward several hundred yards and halted. Here we caught a glimpse of some troops in line to the left of our front, and some of their men began to fire at them. Colonel Archer put an instant stop to the firing, being uncertain whether the parties aimed at us were friends or foes, and ordered me to take the first platoon of Company D and deploy it in skirmish line to the right of the road. Captain Powell, to deploy the second platoon of the same company on the left of the road. Just at this juncture a couple of young fellows came running to us from our left, to tell us that it was Hampton's Legion that was out there. Archer told me that he thought the First Texas was somewhere on our right and cautioned me not to fire on it.

"But the First Texas was not on our right, and I soon ascertained and reported; it was the Ninety-fifth Pennsylvania Bucktails. Archer ordered me to give them h-ll and the boys of my platoon did it, for when formed in line of battle the Fifth was struggling through the undergrowth, it passed and counted forty dead Bucktails, killed by my platoon in its first volley. The undergrowth was so thick that we could see but a little way ahead. While the regiment was thus advancing, I located the whereabouts of the First Texas by the sound of Colonel Rainey's voice, but where it came from the right or left, I am not sure. I did not see him, or the First Texas, but I heard him call out to his men, in the shrill penetrating voice that was so peculiar to him: 'G-d d—n it boys! Swing around there on the right! The Je-e-e-us! Did you hear that bullet?"

The Fifth continued its advance to the river, where the men now pretty well exhausted their long run over and through the undergrowth, laid down in cedar thicket, so near the gunboats that we could see the smoke from their smokestacks, and hear the puffing of the steam from their boilers. The fire of their guns, of course, passed over us.

While lying in the cedar thicket, some troops came up within thirty yards of our right flank, and formed in line facing the river. Think that they were one of the regiments of our brigade, I walked up pretty close to them and asked what regiment it was. 'Eighteenth Ohio' came the reply, clear-cut and distinct. I immediately reported the fact to Colonel Archer, and when he had taken a near enough look at them to see they wore blue, he ordered the regiment to move as quietly as possible to the rear, which they did, and an hour later the Fifth got in line with the other regiments of the brigade."[1]

1. Polley, *Hood's Texas Brigade*, 25-27.

Adjutant Campbell Wood
5th Texas Infantry

DAILY EXPRESS
SEPTEMBER 8, 1907

Flags of Texas

This flag known as the "Wigfall Flag" was carried by the Fifth Texas in the Peninsular campaign, and was in the battle of Eltham's Lading or West Point – the first battle in which the Texas brigade was engaged. The last time I recall seeing it carried by the regiment was in the beginning of the seven days fighting around Richmond, Va. It was then badly soiled, and considerably torn and about this time was exchanged for the new flag made and presented to the Fifth Texas by Mrs. Mary E. Young of Houston, Tex. She sent the flag to the regiment by some of the new recruits who joined us during our retreat from Yorktown in the Peninsula, The Wigfall flag was then sent to our agent Arthur H. Eddy, at the Texas depot in Richmond.

Corporal John N. Henderson was promoted to 3rd Corporal of Company E in May 1862. He was wounded twice that year, first during the fighting around Richmond, then again at Antietam (Sharpsburg,) where his left arm was severely injured, resulting in amputation. He returned to Texas, was meritoriously promoted to lieutenant, and after he recovered from the amputation, he returned to the brigade and served on the staff of General Jerome Robertson during 1864-65. After the war, Henderson became a well-known and respected Judge along the Texas Circuit.

Corporal John N. Henderson
5th Texas Infantry

RICHMOND DISPATCH
JULY 7, 1901

How The Fearful Drama Began

Corporal John N. Henderson, 5th Texas Infantry.

(Tarlton Law Library)

The fearful drama of 1862 is about to begin. In the early spring the Federal army, some 200,000 men, under McClellan, changed its base from the Potomac to the Peninsula at Yorktown of historic memory. They were confronted by Magruder with some 10,000 or 15,000 troops, who held the vast horde of Federal troops at bay until the arrival of General Johnston, who rapidly marched from the line of the Rappahannock to reinforce Magruder. After confronting him for several days, our army began to retreat toward Richmond - Hood's Brigade, then belonging to Whiting's Division, covering the retreat to Williamsburg, passing through that town, while the battle of Williamsburg was in progress. The division was moved rapidly to Eltham's Landing, on York river, in order to cover an anticipated movement calculated to intercept the retreat of the army. Here, for the first time in the campaign, the Texas troops engaged the enemy, in a densely wooded country along the York River. The Fourth and Fifth Texas encountered the enemy in strong force and a sever engagement ensued, in which that regiment drove at least double their number of Federal troops under cover of their gunboats. The entire brigade lost some forty or fifty killed and wounded, while the enemy's loss was at least twice that number. Here it was that Captain Denny of the fifth, and Lieutenant-Colonel Black, of the First, were killed, and Lieutenant-Colonel Rainey of the First, was severely wounded. I mention this battle, not so much an account of its importance as compared with others which ensued, but because it was the first contact the Texas troops as a brigade had with the enemy, and in that engagement it performed its part so well as to received the encomium of General Gustavus W. Smith, the commanding officer. Hear what he says in his first official report: "The brunt of the contest was borne by the Texans, and to them is due the largest share of the honors of the day at Eltham." And again he says: "Had I 40,000 such troops I would undertake a successful invasion of the North.[2]

Sergeant John M. Cox
5th Texas Infantry

GALVESTON DAILY NEWS
JANUARY 31, 1884

The First Man Killed in Hood's Brigade

[To The News.]

Sweet Water, Texas, January 28, 1884. - Knowing that The News has a larger circulation than any paper in the State, that is to read and perused by the many thousands who are more or less interested in our history, induces me to write this article to correct a mistake published at Houston, Texas, and copied by many of the papers throughout the State, about the death of Captain Dennie. The article referred to says that "Captain Dennie was the first man killed in Hood's Texas brigade by the enemy." This is a mistake, and, as trivial as it may be, I want it corrected, Watson, a young man from Leon county, Texas, Company C, Fifth Texas regiment Hood's Texas Brigade was the first man killed - and not Captain Dennie. Watson was killed at Yorktown, and Captain Dennie was killed on the retreat from Yorktown, Va., and on the morning of the day Hood's brigade fought General Franklin's corps at Eltham's Landing; or, as some of our boys have it, at West Point, on the Pamunky River. If my memory serves me right, Watson was killed about two weeks or more before Captain Dennie.

John C. Cox,
Sergeant Company C, Fifth regiment Hood's Texas brigade.

---

2. Excerpt of speech given to Hood's Texas Brigade Association, held in Galveston, Texas on July 6, 1911.

Private Robert Campbell
5th Texas Infantry

We (Whiting's Div) were aroused and placed under arms at 5 A. M. The 1st & 4th Txs – 18th Ga., were immediately marched off and stationed near the point of landing. Whiting's own Brigade commanded by Col Law of Ala. was stationed as a reserve. They had been absent an hour when we (5th Txs) were started on the march. We wen ½ mile and were halted by a "corn crib" and two ears of hard corn issued to each man, a poor nourishment for men who had been two days fasting. We then were double-quickened by Col Archer for nearly a mile and formed a line of battle. We moved in this order into a cape of woods. Our position was about ½ mile above the other troops and we were stationed there to protect the left flank. WE halted but a few moments when we were about faced and double quicked back to "night camp." Had we remained 10 minutes longer we would have been cut off as a body of Yanks had landed first below us and were moving on a line directly between us and the other regts. As we reached our old camp – loud peals of musketry greet our ears – not more than a mile off the battle of "West Point" had begun. Col. Archer ordered us towards the firing at a double quick and after we had advanced a half mile we were halted and the order "load at will" – was given – we were soon loaded – capped – and ready for the fray. We now struck the main road leading to "West Point" – and proceeding down it – to within ½ mile of York River. We took a road leading to the right but running parallel with and ½ mile from river. After moving down this road some 500 yds, the loudest and most continued peels of musketry greeted our ears. The gallant 1st – and brave 4th – with the never-failing 18th Georgia were hard at work at the feast of death. After having proceeded down this road – some 500 yds – and being ignorant of the close proximity of the Yanks – we were startled and surprised by a volley of minnies being poured into the 5th regt – wounding several & killing our commissary a brave & noble man who was riding at the head of the regt. Immediately Col. Archer ordered my company ("A") and Co "E" out on the left flank – as skirmishers, and the rgt drawn up in line of battle. I myself was in a quandary. The old uniform of my comp, was a blue shirt - & in this the primitive days of the revolution the boys had a mania for shooting at anything blue. Major Botts of the regt – came to me and advised be to bid adieu to my old "Bayou City Guard" shirt – and for the sake of life, I set it free – and also bid adieu to a Confederate Grey overcoat – which much impeded my movements. After this sad surprise – we moved towards the river in line of battle hearing all the time the loud musketry of the

1st & 4th hard at work. After having some 600 yards – we halted and formed line of battle at right angle with and 400 yds from the river facing the battle. Directly the Yanks made their appearance – and with right good will did open on them. We whiped them off – killing some 50 – by this time we heard a deafening yell – which we recognized as coming from Texas – we immediately moved forward to the scene of action – and found the 1st & 4th Texas and 18th Georgia – charging and driving them. We joined our left to their right and drove them towards their gunboats – when in a few hundred yds of the river, their gun boats opened on us with grape shot – shell – and canister – Genl Hood ordered us to lay down. It was now about 11 A. M. – We remained in this position until 3 P. M. awaiting a new advance, but they never came. There had Franklin – with his 25,000 men been badly whipped by 7000 determined Rebel – in fact by 3,500 for Whiting's Brigade was not called into action. "Whiting's Brigade" was anxious to come in and join us, but that Genl told them in his blunt, offhand way – "never mind, them darn Texans will eat 'em all up." Genl Hood was everywhere of our Brigade as many as I could learn numbered about 45 to 50. This was our maiden fight and nobly did we sustain the reputation of our State. My emotions on going into battle – admit not of explanation – if reader you have never been where death reigned on the field of battle – you cannot know what feelings move the heart. The 5th Texas was not heavily injured. The "Hampton Legion" of South Carolina under Col Wade Hampton bore a river until a late our at night – a lively chat that was kept up all over camp – each one giving in "his experiences" as for my part, my "prelims" were not the best. I had thrown away my coat and shirt in order to double-quick and now I was cold – in double-quicking after the Yanks my haversack string broke – and now I was minus my "two ears of corn," and for first time contrasted the comforts of home with my present situation. Thankful to my God for his protection – hungry, tired & sleepy – I lay down on my blanket about 11 P. M. Hardly had the delightful and restoring sensations of sleep fallen upon us – when the "long roll" sounded and by 2 A. M. we were under arms. Soon after laying down that night all our trains had left – and as we learnt afterwards – Johnston's veterans from "Williamsburg Field" – had passed by on the road for Richmond. We had saved our army and now must save ourselves and if we had waited for a day we would have been captured. Well – yea very well, do I remember the 2 A. M. – 8th May. Never having been exposed or suffered fatigue – the march of the preceding days – the wetting which I had got was now discovering itself upon me by way of chill and cramps. I was so sock, that I could hardly move – but all of our ambulances had left – and I must either be captured or make an effort. We started at a double quick – and in an hour – I was all right again –

by virtue of the exercise. Had the Yanks known of our retrograde movement they could have injured us much. Strict silence was the order from Gen. Hood, and for 10 miles, not a voice above a whisper was heard, for every moment we expected to be fired into by a Yankee ambush party to show the true state of affairs. I copy from Chaplin Davis (4th Texas) work on Hood's Brigade,

"What a hearty laugh a man could have had, had he been in a position to observe both armies that night. Ours moving swiftly and stealthily along, casting many and anxious glances to the rear fearing to discover the head of a pursuing column. Theirs digging, toiling, and sweating, preparing to receive vigorous onslaught which they knew the Rebels would make at daylight."

We, once more were the rear guard of the Virginia Army. We marched hurried the remainder of the night, and soon after day, we came upon Barksdale's Mississippi Brigade –we continued the march- at an ordinary pace – all necessity for speed having passed away. We continued on the road until 3 P. M. (May 8) when a cluster of houses – a court house, jail – indicated that we were approaching a village – with colors flying and drums beating – we passed through the village – with colors flying and drums beating – we passed through village, which was "New Kent Court House." Directly beyond the courthouse – in an open field we saw Confederate troops - who were able to rest there that night and bring up the rear next day. We marched on some four miles beyond – and struck camp – and a happy moment it was too for four days we had been on a forced march – in the worst weather and on the worst of roads – living on nothing – you might say – had fought a battle and been victorious – and allowed but a few hours rest after the fight. We had been marched hurriedly on and now we had a prospect of a good night's rest and a good and abundant meal. The major part of the army was on its way to Richmond and doubtless near there – our halt here was also to check the enemy if they came on and give the advancing portion of the Confederate Army time to form around Richmond. Abundant rations of bacon & flour were issued us and having cooked our supper – placed pickets on the branch roads to prevent a surprise – we returned to our "virtuous couches." For one who had hardly ever seen cooking done – I took it to finely – except in making my dough. I never could get it to stick together and at first I invariably made my biscuits so hard that they had defied mastication.

Next morning (May 9th) we were up by ... we rested on our arms until 2 P. M. – when the "long roll" beat – as yet the Yanks had not made their appearance. By half past 2 P. M. we took up the line of march for Richmond. Soon after starting a heavy rain began to pour down – and now indeed began the "tug of war." By dark – with the present and preceding rain, the road became almost impassible and by 7 P. M. we could distinguish nothing five feet before us – so impenetrable was the darkness – but necessity compelled us on – and we thwarted the storm. We continued the march until 3 A. M. – passing over broken down and deserted wagons, mules and played out soldiers. Despite the hardship of the march – many ludicrous things occurred. "Kerplash" – could now and then be heard from ahead and behind. Then would follow a big laugh by many mouths and some fellows would cry out – "come here Love & I will pick you up" – my humble self – rolled into a big ditch of mud and water – and was congratulating myself that my eyes were safe. By 3 A. M. – we came upon what is known as "confusion – with ah hundred voices hollering and cursing their wearied mule teams. This proved to be "Chickahominy Bridge" over the Chickahominy River – a stream which from the scenes transpiring in its vicinity is known to history and to "Song."

Then hast we marched since 3 P. M. to 3 A. M. – 12 hours – and had made but 6 miles – I again quote from Parson Davis's.

"Here we found several Genls. all exhorting us to 'close up' and for God's sake to hurry up and cross the bridge."- this was done as soon as the road way cleared. After crossing "Chickahominy Bridge" – we moved on and went into camp. By 5 A. M. the rear guard, had crossed the bridge, and the pioneers went to work – and soon that bridge was among the things that were. We went into camp by 4 A. M., built fires, hung our saturated clothes out to dry – received a cup full of flour ¼ lb. of bacon for our next day's rations. We were as much worn out men as you could find – but nature demanded sustenance, and soon the frying pan and skillet were performing their offices. As for myself I concluded to feast nature on a few "flap jacks" and a speck of bacon. I immediately hid my bacon and three flap jacks – and taking the remaining one – I deposited it in my haversack for next day's sustenance. We lay in camp until 11 A. M. next day – and then took up the line of march for Richmond. By 5 P. M. we found ourselves in 6 miles of Richmond near the "nine mile road" – and went into camp in a grove – near which lay some 40 cords of wood – and in 10 minutes not a stick was left – so soon as the companies had taken their camp ground. A rush was made for the wood – and each fellow busied itself – obtaining wood for his "mess." That night ample rations were issued – and we feasted on "flap jack" and "slush" – two army dishes – which admit of no description. At 9 A. M. (May 11th) we obeyed the "long roll" & started on the march and by 2 P. M. we were in camp at "Pine Island" – a beautiful pine grove on the Richmond & York River RR –

and two miles from Richmond - with a view commanding the whole city.

That was the retreat from Yorktown performed one of the most brilliant military movements on record. Genl. Johnston naturally retreated without any loss - but fought & won two battles - "Williamsburg" & "West Point" - captured many prisoners - and many arms. He was pursued by a Yankee army three miles his number, well-disciplined and equipped. I am by no means an admirer of "Old Joe" - but I believe in the adage "Render unto Caesar, that which is Caesar's due." For this brilliant move Genl Johnston deserves the gratitude of his country. For this single move - he deserves the eulogies of his countrymen - as a move - brilliant in the extreme.[3]

Private Arthur H. Edey
5th Texas Infantry

THE STATE GAZETTE
MAY 24, 1862

Letter from Richmond.
The following letter we take from the Telegraph.
Office Texas Brigade.

Richmond, April 30th. 1862

Editor Telegraph - Capt. Sterret leaves for home tomorrow, and as the mail facilities between camp and Texas are suspended, I thought I would be best serving the Regiment by giving a synopsis of affairs here to your readers in Texas.

The Brigade (1st, 4th, and 5th Regiment,) is within two miles of Yorktown. Health generally good. Those that are sick come here.

The entrenchments of either party are but 150 yards apart. Each relief of men goes to the pits at night and remain three days.- Whenever any portion of the body or head is exposed above the ground, bang! Goes a rifle. So far, two have been wounded in the 1st, the same in the 4th, and one, Wm. Watron, of Leon county, killed in

3. Stoch, George and Perkins, Mark (eds.) *Lone Star Confederate: A Gallant and Good Soldier of the Fifth Texas Infantry*, College Station, TX: Texas A&M University Press, 2003, 11-17. The pen name, Joe Joskins, is believed to be Private Campbell. He was wounded in 1864, and in recuperation wrote his diary under the pen name of Joe Joskins.

the 5th. Before the Brigade was stationed at that point, the enemy was very bold; walking on top of the entrenchments and occasionally drilling in front of them because they felt secure, the regiments of Georgia and North Carolina having only the common musket; but now that the Enfield rifle has been brought to bear, they stand aside. They were fond of climbing trees and shooting down into our pits, but that has been stopped by the Texians. There are some 80 scouts from the Brigade, who hover about the enemy and annoy him as much as possible; one of these, a Mr. Templeton, of the Polk County Rifles, the same who so distinguished himself in the fight on the Occoquan, one of the brave eight, visited and conversed with Gen. McClellan. This I have from one of the boys just down from camps. This same man has killed some ten or fifteen of the enemy, including a Major.

A very heavy battle is expected very soon, perhaps two - one at Yorktown and the other near Fredericksburg. The greatest confidence exists in the army as to the success of our arms. The loss of cities on the seacoast is expected, but on land the South feels her superiority.

All that country in Virginia which has lately been so familiar to your readers, is now in the hands of the enemy, and, sad to relate, much of our baggage and stores.

But we look for a change in this aspect of affairs. Some recruits are arriving rapidly, and the army is increasing day by day.

If anything of interest occurs, I will take pleasure in writing to you, and I would suggest that parties sending letters, &c., for the Regiment, would consign them to the Texas Deport, and we will see that they are sent to the Regiment.

Yours very respectfully,

Arthur H. Edey,
Agt. 5th Regt Texas Volunteers.[4]

Joined Company F, Fifth Texas Regiment in Richmond, Virginia. He fought at Chickahominy, Seven Pines, and around Richmond, served as the commanding general's scout at the battle of Gaines' Mill,

4. Arthur H. Edey enlisted as a Private in Co. A, 5th Texas Infantry, on July 19, 1861, in Houston. He was detailed as an agent for the regiment in Richmond on Feb. 7, 1862, and, in that capacity provided mail service between members of the regiment serving east of the Mississippi and their correspondents back home. He was wounded and captured at The Battle of Gettysburg on July 2, 1863, and sent to Fort Wood, in New York Harbor. He was paroled on April 15, 1865.

and was wounded at the Second battle of Manassas. He returned to the front, took part in the battle of Fredericksburg, and was made a temporary corporal. In 1863 he fought with Robert E. Lee's army and with James Longstreet's corps at the Round Tops and at Gettysburg. Under Joseph E. Johnston he was wounded again at Chickamauga. In 1864 he joined Company E of the Eighth Texas Cavalry (Terry's Texas Rangers) and accompanied the unit on campaigns in Tennessee and Georgia.

Private William A. Fletcher, 5th Texas Infantry.

(Texas Heritage Museum Historical Research Center)

Private William A. Fletcher
5th Texas Infantry

I later joined my command near Yorktown. We were now under command of General Magruder, and here began my service on the front. Magruder had fortified part of his front by putting a levy across a piece of woodland, boggy, flat. This was partly filled with water and made quite a stretch of front that was easily guarded. There was part of a battery stationed to protect it. My first sight of the enemy was crossing the levy and deploying some distance to the front with a large detail. Toops and I were the detail of Company "F," and he and I were in position on an outpost, near center of the line, and almost one hundred and fifty yards apart. Our line, from the best I could observe had both right and left resting on swamp, and was formed as part of a circle. The position was in the woods, with no opening in sight. After remaining in position some time I grew restless and felt that we would not have a chance to see a Yankee and from my early education I was satisfied they could take a challenge and not resent it on fair terms; for there we were in their line and ready to fight and as they were invaders and hunting a scrap, they were cowards or they would not take a dare. All such foolish ideas as stated, had been thought of a number of times, but along well in the turn of the evening, when my patience was near exhausted, I heard a gunshot to my right and it was quickly followed by others and nearing my position, I was all eyes to the front and before I was aware of it the land had given away to both right and left, so the first shot that I fired attracted fire to my point. I soon saw that I was about to be cut off, so I turned and put in my best licks for the levy, slackening speed now and then to reload; but at each time the whizzing of a few near bullets said "Faster." Some of the Yankees were good runners and by my slackening to load, had gained on me. When I got loaded I dropped behind a log and started to shoot ---- the bullets were whizzing from front and both sides, and I saw my only chance was to run, for if they could not hit me, they would soon catch me; and from that point to the levy I tore through the brush and over logs, making such a noise that I heard but few bullets. When I came near the foot of the levy I saw the most of the men crossing ---- there seemed to be no one in charge and were all panic-stricken. Under the impulse of the moment I called "Halt and hold the levy." All the men who were near me about faced and commenced firing; and a soon as the men on the levy heard the command to halt and the firing of those who had halted, they returned, I think, without an exception. We were well protected under the hill by good size trees and we fired at least fifteen minutes before the enemy gave way, but some of them had gotten within less than one hundred yards of us. The enemy were shooting what we supposed to be 'explosive bullets," from the noise they made when hitting a tree. I got several shots, but I did not think any of them hit the mark and I soon satisfied myself from the way the enemy could get from one tree to another that they were expert woodsmen and called Toop's attention to it and compared them as reminding me of wild turkeys, I don't think we killed or wounded one, and whilst they made a number of close shots, there were not one of our men hurt.

On returning to camp my remarks were contrary to my education. My words were "Boys, the impression that we have about the Yanks being poor woodsmen and marksmen will not hold good, if our levy experience is a sample; from a small boy up; I have had the association and advice of both the white man and the Indian in Woodcraft, and I think the Yanks that we fought were as expert getting from one tree or log to the other as ever I saw, and they reminded me in cunningness of wild turkeys; so I think now, if we wing one a piece through our service, we have done a good job, and the thing that now interests me most is to find out where those Yanks are from, and if there are many of them; so boys, let me hear from you, if you hear anything about them. "I suppose there was some interest in the matter, for in a few days I was informed that they were "Western trappers and Hunters." My reply was: "Thank God for that; for the great odds we will be forced to fight, there would be but few of them."

Toops was an old acquaintance, and possibly a distant relative of Captain O'Brien, and was in the captain's mess. The captain was a noble specimen of humanity; was very brave, just and kind and open to expression, but thought himself a correct judge of men. It was his hobby that if he had a regiment of such men, calling the names of a number that were in the company, he could accomplish anything in the way of battle that could be done by the same number of men. His idea of a soldier was the wild, reckless, camp-fighter, the ones the "slowpokes" would have to guard when in the guard house; so I with quite a number of others was classed as an all-around good camp-man and would fight in battle line with the others to lead. It was not long after we returned to camp before one of our boys said to me: "Bill, I heard Toops tell the captain that he had thought him a good judge of a soldier, but now he questioned it and he thought before we got through he would be of the same opinion."

As we were camped some distance to the rear, we heard but little from the front, so we passed quite a time in drilling and camp duties, with plenty of rations, but vegetable food was very scarce, less wild onions; and I think they were the most plentiful and the largest I ever saw, but one hearty meal of boiled bacon and wild onions will satisfy one's craving for vegetables for some time.

In the course of time, Magruder's forces commenced to move ---- of course, we knew nothing of the why and wherefore of such moves; but it was only a short time until we had a brush with the enemy. There was considerable rifle practice by both sides, but little damage was done. The enemy shelled the woods liberally with what the "knowing ones" called "mortar guns" from their boats. I know the shells were large and made a frightful sound as they passed over us and exploded beyond. The 1st Texas Regiment had a company with quite a number of Indians enlisted and from what I learned, they protested against such warfare, as the fellow who shot those big guns was out of reach of their rifles, and they were not having an equal show. I think this ended the Indians' service, as I understand they were sent back home.[5]

---

5. Fletcher, William, Rebel Private, *Front and Rear Experiences And Observations From The Early Fifties Through The Civil War*, Beaumont, TX: book, 1905, 17-20.

Private John M. Jenkins
5th Texas Infantry

HALLETSVILLE HERALD
JULY 16, 1896

He was never a robust man, and for many years had suffered with rheumatism and other troubles. Despite physical defects, he enlisted in 1861 in Co. B, 5th Texas regiment, Hood's brigade, and was wounded in the battle of West Point or Eltham's Landing, his watch saving his life. He was shortly after discharged for disability and came home.

Private Joe Joskins
5th Texas Infantry

Our division (Whiting's) was aroused from its slumbers at 12 am on the night of 4 May, and started on march. The rain pouring down in torrents, and so dark we could distinguish nothing. We continued in the march the rest of the night and during the days of the 5th and 6th. The rain never ceased, mud & water knee deep, and misery & fatigue. Was viable upon the countenance of all. The rain cleared during the evening of the 6th and we went into camp, and spent a most delightful night. Early in the morning of the 7th of May, we were up again, we resumed march, not toward Richmond as all reported, but towards West Pt. We marched but a short distance, where we were halted. The 5th Txs was sent out on a reconnaissance. We had gone but a half mile, when we came back to the position in a hurry, having fallen into a Yankee trap, but the spring being rusty, did not work quick enough to ketch us. On returning to the place where we had left the rest of the "Brig" we found them gone. A corn crib being near, the commissary issued two ears of corn to each man, which was quite acceptable, since we were out of rations. Corn is certainly very substantial. We had hardly shaved the corn into our haversacks, when the crack of the rifle and the loud boom of cannon, was heard, not very far off. We were quickly formed in line, and at a double quick, started to the rescue of the other regts, for we were certain they had found a mouse. We were pushing ahead on the West Pt road. My company (A), and Co "E" were thrown out on the flanks of the 5th Texas, very suddenly and unexpectedly, we were greeted by a volley of "minnies" whose music, though sweet, is more dangerous than that of mosquitoes. Several were wounded and Capt. Denny, commissary of the 5th who was riding at the head of the regt, fell dead, as a matter of course our first consternation for the moment, possessed the sould of each, and as we had time, to collect our nerves, the deep and heavy roll of musketry, the long, loud and terrible crash of the "cannon ball," as it came whirling by us, told us we were about to be realized in the battle of Eltham's Landing.

We soon rallied, formed and took our position at the fork of two roads, directly after which we were guided by the presence of the Yanks, whom we soon had running, killing many of them on their expended Southern forms. During this time the 1st Texas, was bearing the brunt of battle under the gallant Col. Rainey. They were charged and charged again, each time delaying the Yanks, as each charge was ordered, and with the 4th, 5th and 18th Ga, we soon had the Yanks fleeing for dear life. Leaving guns, clothes & everything a "reb" could desire, killing them at every stage. We ran them back to their Gun boats, near which we took a position, and lay for three hours, under a very severe gun boat fire of grape and canister. Thus did the "Texas Brigade," numbering some 35,000 men, whip & drive back a Yankee corps (Franklin's) of 20,000 which had they been brave men, and not mercenary soldiers, and had they been fighting in a righteous cause, might have destroyed our army. Texas need not blush at the work of her sons, on 7th May 62. The loss of the whole brigade was 40 odd. The 1st Texas losing more than any other. The Yanks showing no disposition to again try their experiment of cutting of an army. We left our close proximity of their gun boats, and returned to our camp, which we occupied before the battle, and till dark the camp reminded one of the noise made in a pond of frogs, every individual giving his experience, the emotions he had, and the part he played in his first battle. This being our first fight, we of course thought it was a "Waterloo." In the year 64, it would rank as a slight skirmish. Major Genl G. W. Smith of the "Virginia Army" thus speaks of us, "The Texans have won immortal honors for themselves, their state, and their commander Genl Hood, at the battle of Eltham's Landing." But in praise of the "Texas Brigade" I could talk a week and then not say half of what they deserve. Genl Melton thus speaks, "Here we had a fair example of your Texans, under Hood. The best fighters in the Confederacy, men on whom we could depend, who seem to fight fir the very love of it." While the boys were all giving in their experiences, I felt rather gloomy, I was tired and hungry. I had no remedy for the latter, for in the morning, I had thrown away my overcoat in order to double quick, and in double quicking I had lost my haversack, containing 2 ears hard corn, 1 cracker, and 3 bites of bacon, so my feelings can be better imagined than experienced, besides this, I was seized with a fever & cramps, and in the language of "Bowers" wished myself most dead. We lay down to rest that night, flushed with victory, and proud of our 1st days' work, and were just on the eve of dreaming of home and the loved ones there, when the long roll aroused us from our slumber just begun, and the order "prepare to march" was passed from mouth to mouth. We saved our army and now we must save ourselves and should we wait for the day, our situation would be iatrical, since the whole army passed on its way to Richmond.[6]

Private William Henry Matthews
5th Texas Infantry

POLK COUNTY ENTERPRISE (LIVINGSTON, TX)
OCTOBER 7, 1984

A private soldier sees nothing except what is just before him and knows nothing of what is going on elsewhere. My first introduction to the Yankee bullets was at the little fight of Eltham's Landing, where our regiment was in advance of the Brigade marching along an old road in ranks, when suddenly the head of the regiment was fired into by a squad of Yankees lying in ambush. Our Quartermaster, Capt. Denny, was shot through the brain and fell dead from his horse. Someone else, I think was wounded. I have always believed that Capt. Denny was the first man killed in Hood's Texas Brigade.[7]

P. P.

"Nom de plume"
5th Texas Infantry

---

6. Joskins, Joe, *A sketch of Hood's Texas Brigade of the Virginia Army by Joe Joskins A rebel in Co "A" 5th Texas Vols "Hoods Texas Brigade, "Fields Division, "Longstreet's Corps," Army of Northern Virginia, June 18, 1865*, UTSA Digital Libraries, Rare books, collection, July 18, 2012. "Joe Joskins" is the Nom de Plume, for Private Robert Campbell.
7. Private William Henry Matthews. He fought in all the battles with the 5th Texas except the battle of Gaines' Mill, leading up to Gettysburg, where he was wounded in the arm and captured. He was imprisoned at Fort Delaware. His brother, T. C., who was also captured at Gettysburg, was killed by a Union guard who previously was a Confederate soldier, but switched sides.

THE TRI-WEEKLY TELEGRAPH
JUNE 27, 1862

The Battle of Chickahominy
Special Correspondence of the Telegraph.

Line of Defense
Near Richmond, June 3d, 1862.

Friend Cushing - The long looked for battle has at last commenced. An unmistakable index to the general engagement was given forth on Saturday, the 31st inst. The prologue was spoken by our scouts and the first chapter that followed was full of interesting details and intensely exciting incidents. It came at an hour when totally unlooked for as thought by the knowing ones that a battle was impossible, owing to the unfavorable state of the ground, caused by the heavy rains a few days previous. Fate decreed it otherwise, however, and the anticipated event was ushered in to the surprise of all. The conflict though, did not find the citizens unprepared for the exigencies of the occasion. For the past few days everything has been done to render the accommodations for the wounded equivalent to the demand for attention. Churches, public buildings, commodious stores, private dwellings, in fact, every place that could be spared n a case of extreme necessity was freely thrown open for the reception of the defenders of the soil. Meetings of the citizens has been called and committees appointed to collect buggies, carriages, and other spring vehicles of every description, to carry the wounded from the field.

Medical stores were on hand in abundant quantities, and any number of surgeons, army and resident, were present to allay the fearful suffering of the brave men, who had fallen a prey to the balls of the Yankees. The patriotism evinced by a majority of the citizens, was unparalleled, and too much cannot be said in their favor.

The Battle

Commenced early on the morning of the 31st, and raged with unabated fury during the entire day; the skirmishing began about 9 A. M., and the general engagement was not deferred long thereafter. The division commanded by General Longstreet was brought up with bands playing, colors flying, and the men in the best of spirits; they proceeded immediately to the attack with all the hilarious delight,, that might have been appropriate at a wedding feast; the road they had to pursue was almost wholly impassable, owing to the terrific thunderstorm of the previous day. Nevertheless they pushed on, nothing daunted by the seemingly impassable barrier, to the field of strife; the enemy had anticipated the attack, and had been very active in their preparations for it. At about 11 A. M. the work of slaughter commenced in dead earnest, but under circumstances very unfavorable to our men. The opening charge was made by the 28th Georgia, 2d Mississippi and 4th North Carolina regiments and was probably the most desperate onset of the war. The enemy in superior force were entrenched strongly, having a swamp in front of them running parallel with the Chickahominy; during the night they had felled the trees, and stationed their sharpshooters all through the swamp, at every available point, and so closely were they concealed that not a sign of an enemy in person could be seen, though our forces had marched within 50 yards of the foe; our ranks in the meantime having been literally decimated; and still suffering the effects of a heavy fire.

The order came through to "charge," and our men, utterly regardless of consequences, rushed forward in gallant style, the leaden shower still pouring into them. Near one half in the charge were placed "hors du combat," without having so much as seen an enemy. The remainder boldly pushed on, driving the sharpshooters back to their breastworks, in a rather precipitate manner; then following them up they charged the trenches, fighting with an utter disregard of fear, scattering the foe, and driving them rapidly in front of them. Brigades of the enemy were now sent up and re-attacked our men, and the three gallant though thinned regiments step by step were being forced back. At this critical juncture, a brigade of Louisiana troops came up, and again "forward" was the command. "Remember Butler," was the cry as through the rain of ball and grape our men marched a second time to the breastworks, and notwithstanding the great odds, drove the enemy away, causing many to bite the dust; the hottest firing now commenced; the whole of Longstreet's division having been brought up, and engaged, regiment after regiment, the enemy were driven from their hiding holes, fighting desperately, but still forced to give way, our forces still following them, and driving them from every accessible point, capturing guns, fortifications, tents and supplies of all descriptions, in abundant quantities, and driving them pell mell beyond the original point of attack two miles.

They left hundreds of killed and wounded on the field, besides provisions, ambulances wagons, horses, and one hundred barrels of whiskey. – Routed and driven from their entrenchments they made another effort to regain their lost fortunes by that most potent of all

Yankee maneuvers, a flank movement; but the ever watchful Johnston was awake to their intentions and their force was met by a portion of Whiting's division composed of Mississippians, who without waiting for the enemy to come up, attacked them, drove in their skirmishers, charged the main body and drove them into a dense swamp, capturing two masked batteries they had erected preparatory for a retreat. The fight was somewhat lengthy, and darkness had fairly set in by the time the enemy were driven back. When night came on the enemy had been ingloriously defeated in two attacks, driven three miles beyond their encampments, lost fifteen guns, an immense quantity of stores, and at least 3000 killed and wounded.

This statement of the "Federal Victory" of Saturday and they enjoyed it undisturbed during the night having retired three miles; while the "poor discomfited rebels" had to put up with the rough fare that had fallen into their hands and for the first time in some months we were compelled to drink strong coffee and fine wines and eat those vile things called sardines, pickles, preserves and cracknel biscuit. Truly, such defeats will ruin our army; gouty symptoms have already been exhibited.

The distant roar of cannon was the only thing that disturbed our equanimity during the night, and at an early hour Sunday morning, hostilities were again resumed, the enemy manfully attempting to regain their lost ground, but they were gallantly repulsed by Huger's division. The fight lasted a greater portion of Sunday, our forces driving the enemy back gradually. Some 500 prisoners were captured this day, and the loss on both sides was very heavy. The enemy retired into the recesses of the swamp, and the fighting for the day ceased. During the balance of the day the operations were not of a very important nature - both parties preparing for the next grand attack, which it is supposed, will take place either Monday or Tuesday.

The Texas Brigade, during the entire fight were marching through swamps, always on the qui vie to render assistance, but never having an opportunity to join in the combat, though constantly exposed to a scorching fire of bombs. It was very tantalizing being on the reserve - balls striking around promiscuously, and no chance to reciprocate the favors. Our scouts, however, a detail from three from each company, were more fortunate and had many fine opportunities of displaying their prowess, which you may rest assured they did to great advantage, never letting an occasion pass to render good and effective service. They harassed the enemy considerably, and captured many prisoners; and, I am happy to add, they all escaped without a scratch. They were all complimented by Gen. Hill for their gallant conduct.

The boys were fearfully musical last night, having captured the instruments of three brass bands. Our men have in their possession any quantity of Yankee letters and stationary. It seemed as though each regiment had its private stationary, with the picture of its Colonel and the Union coat of arms stamped on the paper and envelopes. Among the spoils were several late Northern papers, a Herald of the 29th and Frank Leslie's 31st, copies of which I send you.

Our Wounded

Have been constantly pouring into the city since the fight began. Yesterday evening the city presented one of those grand spectacles which it is not often the province of man to witness. Hundreds of wounded were brought in, and were immediately taken in charge by the ladies, whose faces betokened the finest sense of sympathy; nor did their actions denote less than their faces. Some dressed the wounds, some fed the hungry, some with soothing cordials cooled the parched lips, while their sweet forms were ever present, and with saddened hearts but cheerful faces administered those little delicacies, which revived in the soldier's breast that hopeful remembrance of their homes, in whose defense they had so nobly fought and fallen.

On the battlefield the scene was terrible in the extreme. Pen cannot describe, nor imagination conceive, the picture that presented itself on every side. Friend and foe were scattered here and there in death or in their last agonies; far and wide are deserted camps; dead and dying fill the tents; horses wounded and lame were rushing to and fro; and every species of wound that a mind can conceive, or known to the human body could be seen in ghastly reality. Water was the only cry, and the blood stained slimy floods were eagerly drank up by the men; sighs and piteous moans could be heard throughout the entire field, some with eyes up-turned uttered an humble prayer for mercy, while others with blasphemy on their lips, spoke nothing but imprecations. The wounded may be numbered by thousands. We lost two Generals, Hatton and Pettigrew. General J. E. Johnson was wounded. A list of killed and wounded so far as known, can be compiled from the Richmond papers which I send. The prisoners taken nearly all expressed themselves as being glad to get out of it, and they manifested considerable good feeling towards our men, laughing, joking, complimenting, etc.

The result was very favorable in the fight of both days. We drove the enemy from all points where we attacked him, and took 29 pieces of cannon, with 850 prisoners.

In this glorious contest every State of the Confederacy was represented, and the sons of each strove to add new luster to their fame, and render their deeds the subject of another page in their country's history. The grand finale is expected tomorrow, when I trust I may be able to write again and chronicle all safe. Hoping that Providence will aid us in the continuation of the struggle,

I remain yours, &c.,

P. P.

N. B. The following private note accompanies the above letter. It was not intended for publication, but it will interest many. The Dempsy Walker mentioned is one of the representatives to the Telegraph office in the Bayou City Guards. We are glad to hear so good an account of him. The name of the heiress is still in reserve for the bravest of the Brigade. We can tell our "special," however, that it will take something better than dodging bombs to get it.

Private Riley Luther Scherer
5th Texas Infantry

'Good Bye Dear Sister' The Unhappy War Of A Texans Who Was the First To Enlist And The Last To Fall, by William A. Palmer Jr.

Riley Scherer was the son, grandson and nephew of Lutheran pastors. A decade before the war his father, Gideon, had left his pulpit in Virginia and moved the family to the Texas frontier to minister the grown number of German Lutheran immigrants to the Lone Star State. Gideon and his brother John Jacob would also establish Colorado College in Columbus.

In addition to ranching in Columbus, Riley taught school before the war. He never married, and seems to have been affected by his widowed father's death in June 1861. He spent the next several months acting as co-executor, along with his uncle, in settling his dad's estate.

By the spring of 1862, the 5th Texas Infantry, part of John Bell Hood's Texas Brigade, was in desperate need of replacements. The regiment had suffered greatly from illness during the previous winter near Dumfries, Va. When recruiters from the 5th arrived in Columbus in March 1862, 27-year-old Riley Scherer volunteered.

By the time he reached camp in Houston, however, the former schoolteacher was already having second thoughts. "We have some bad fellows among us," he confided to his sister. Even before the 37 men who joined up in Columbus had left town, new enlistee William Ryan was stabbed to death, reportedly "by a man named John A. Pierce."

Riley flossed his letter with a benediction and a confession: "May the Lord bless and keep you from harm. I will be without a friend all the way to Va." His sense of obligation and foreboding clearly worsened during 18 days of trips between Houston and Richmond, Va. Hungry and exhausted, he admitted to Mary Ellen:

> **"This morning I had almost given away to despondency, I want to be home, there is no place like home. If I had a good Christian friend with me or even one that respected morals I would be much happier. But O the wickedness, the wickedness that is carried on. Imajenation cannot paint the black heartedness & the desperate wickedness that is going on....I feel so solitary & lonesome. Sometimes I get about half asleep and imajine that some of you stepped in and fell upon my neck. I cannot describe the joy I felt...I jumped up on the bench and wakened and say my sad disappointment. I do not know if any the boys seen me jump or not."**

Scherer did not share the optimism of some of his comrades, writing, "I still entertain the same opinion about the war as I had before I left. I still cannot think that the South will be subjugated." Despite being homesick, Scherer was determined to honor his soldierly obligations. He wrote:

> **"Our regiment is at Yorktown about 60 m. from Richmond....Tomorrow I think we will start to our regiment. Every body says that there will be a desperate fight here or at Fredericksburg in a few days. We will have to go right into it probably as soon as we get there....We will have to stand up to the rack hay or no hay....Sister you may get this letter and it may be the last one God only knows. Before you have time to get this probably I will be a wounded soldier or a corps upon the field of blood. I have given myself and body up into the hands of God and feel satisfied that he will do what is best for me. If it is his will that I should be numbered among the slain, I say his will be done not mine.**

In that same letter, he also sent a message to his aunt that offers a glimmer of humor, a clue that in other circumstances our picture of him might be far different: "Tell Aunt Leah that I would much rather ride over the prairie after cows." But his final words confirm Scherer's gloomy outlook:

> **"Every Friday I spend particularly in prayer and supplication to God for you, that is you 4 girls. I hope you will not forget me....Good bye dear Sister God bless you all, and if we never meet again on earth again I hope we all will see each other in heaven. And now dear Sister do not forget your Affectionate Brother Riley.**

Scherer evidently shared a premonition that he would be wounded or die in combat with others in his unit. James Daniel Roberdeau, a former lieutenant in his company, later recalled that several comrades had presentiments that proved to be uncannily accurate. After relating a story about a soldier killed at Second Manassas, Roberdeau noted that man was the fourth he knew to have made similar predictions: "Denny and Scherer [sic] at Eltham's Landing, Gaines' at 'Gaines Farm,' and Pinchback at Second Manassas."

Scherer and his fellow recruits had joined the 5th Texas at Yorktown just as General Joseph Johnston was issuing an order for the withdrawal of Confederate forces. On May 5, Union forces caught up with the Southerners just east of Williamsburg. A confused fight in a driving rainstorm bought Johnston's forces time to continue their withdrawal.

Even while the fighting was raging near Williamsburg, Maj. Gen. George McClellan was supervising the loading of Brig. Gen. William Franklin's division onto transports at Yorktown. His plan was to undertake an amphibious landing near West Point, at the head of the York River, at which point Franklin's division could fall upon the retreating Confederates' flank. Since there were insufficient transports to move the entire division in a single trip, Franklin was ordered to establish a bridgehead and wait there until all his forces arrived.

Johnston, however, was aware that McClellan might make such a move. Early on the afternoon of May 6, as Franklin's first wave began disembarking on the south bank of the York, the Southern commander ordered the division of Brig. Gen. W. H. C. Whiting, along with the infantry component of Colonel Wade Hampton's Legion, to interpose their commands between the Confederate line of withdrawal and the Union bridgehead.

The Union bridgehead around Brickhouse Landing was certainly defensible. Mill Creek ran from left to right in front of Franklin's semicircular bridgehead before emptying into the York River upstream from the landing. Bakers Creek protected Franklin's left. Offering an additional measure of protection, Mill Creek was impounded by two dams that created substantial millponds on Franklin's front. Bassetts

Pond, the larger body of water downstream, was named for the colonial owners of nearby Eltham Plantation.

Narrow roads traversed both dams and offered the only access to the bridgehead area on Franklin's front and right. On Franklin's left, the main road between Brickhouse Landing and the village of Barhamsville, about four miles away, skirted the headwaters of Bakers Creek. A Barhamsville this road intersected with the road between Williamsburg and New Kent Court House, which the Confederates were using for their withdrawal.

The Confederates took the initiative. Whiting ordered Hampton's Legion, which was at Barhamsville, and the Texas Brigade, in bivouac just north of that village, to advance on the Union bridgehead. They moved out at daybreak on May 7, just as the final elements of Franklin's force were arriving at Brickhouse Landing.

Because both Hood and Hampton would be approaching the Union position through poorly mapped and densely wooded terrain, the Confederate commanders were concerned about the potential of friendly fire. Hood instructed the Texas Brigade to advance with unloaded muskets.

Hood also decided to divide his force. Leaving the 18th Georgia and brigade artillery in reserve, he led the 4th and 1st Texas toward Franklin's center. The 5th Texas was sent along a "blind road' that would intersect with the main road from Barhamsville to Brickhouse Landing. They would link up there with Hampton's Legion.

A daybreak two companies of the 16th New York passed over the dam impounding the upstream pond on Mill Creek and entered what was described as a "stumpy field" where pine trees had recently been harvested. Near the Robert Timberlake House they ran into the 4th and 5th Texas, most of whom were carrying unloaded weapons. In the ensuing exchange Hood, riding at the head of his column, narrowly escaped being killed when fired on by a New York soldier. But one 4th Texas soldier – who had disobeyed orders and loaded his weapon – quickly dispatched the Federal shooter.

Around that same time the 95th Pennsylvania was moving toward Barhamsville via the main road. As the Federals emerged from a densely wooded stretch of the highway into a cleared area, the regiment formed a line of battle, with one battalion on either side of the road.

Colonel James J. Archer of the 5th Texas was concerned about connecting with Hampton's Legion as he approached the main road from Barhamsville – as was Colonel Hampton, at the head of the unit bearing his name. Hampton subsequently reported that "Fearing a collision between my command and that of General Hood, should they be thrown together in the deep woods. I acceded to the request of Colonel Archer, and allowed him to proceed me...."Thus the Texans reached the intersection first, to find the 95th Pennsylvania waiting. Lieutenant Campbell Wood the Texans' adjutant later recalled:

> **"We marched down the road...in column of four ranks. Colonel Archer and Captain Denny rode at the head of the regiment, and I trudged along on foot, immediately behind Archer. Archer and Denny rode slowly, and the men kept close up, talking a little as was usual on the march. Just as we approached quite near to an old shack of a house, a federal sergeant and eight men jumped from behind it and fired a volley at us. All their bullets excepting one went wild, but that one struck and killed Captain Denny."**

Shocked by the death of a fellow officer right in front of him, Wood probably did not notice that the Pennsylvanians' Lorenz rifle-muskets had actually exacted a greater toll. Others would report that Captain W. D. Denny, the company's original first sergeant who had been promoted and transferred to regimental headquarters, was not the only man lost in that fusillade. Riley Scherer also lay dead in the mud of the "blind road."

Ironically Scherer, the last man to enlist in his company, had been the first to die. He and Denny were among the dozen Confederates lost that day as part of the Battle of Eltham's Landing, all of whom would be buried at or close to where they fell.

General Hood reported that "at 2:30 P.M., he "gathered up the killed and wounded in perfect order to the bivouac I left in the morning." Because Hood was with the 1st and 4th Texas regiments throughout the battle, it seems likely he is referring to the dead and wounded of those regiments. Their remains were placed in a common grave north of Barhamsville, where the Texas Brigade had camped the night before the battle. But according to locals the bodies of Scherer and Denny, who had been killed on another part of the field, might have been buried elsewhere.

Scherer's siblings and relatives would later try to figure out exactly where his remains lay, questioning surviving members of the 5th in the decades after the war. In 1915, for example, Scherer's nephew, Jacob Wilson Wirtz, began writing to Allen Harrison Carter, who had served with his uncle in Company B. Carter, then 73, demonstrated his remarkable memory of the May 7 incident – though he could not resolve the family's most pressing question:

I was in a few steps of Riley Scherer when he was shot but we passed him into the fight and never came back that way. I do not know where he was buried. Generally there was a detail made to bury the dead where they fell [.] If I ever knew where he was buried I have forgotten...There was two killed that morning Doc Denny and Riley Scherer... I do not remember the name of the roads [.] I remember the place distinctly. I will make a little diagram on the back of this page.

Carter's diagram accurately depicts the intersection where the 5th Texas collided with the 95th Pennsylvania. It also shows the Texans' line of march along the blind road, which Carter labeled "just a country road," that intersects on the perpendicular with the Barhamsville to Brickhouse Road, about which he wrote, "this was a larger road." Across an intersection he drew an "old field" and at its back an "old house," no doubt Lieutenant Wood's "old shack of a house," where some of the Pennsylvania opened fire. He also drew arrows to show fire coming from the Texans' left, where the 95th Pennsylvania battalion on the near side of the road would have been in a position to fire into the Texans' flank. If Denny, at the head of the column, was hit by fire coming from close to the old house, the arrows on Carter's map suggest that Scherer was killed by flanking fire from the left. Carter wrote at the bottom of his map: "Where the cross is about where they were killed a short distance from the cross road [.] Cannot remember the distance probably one or two hundred yards.

He road on Carter' map still exist today. They are paved, but little else has changed. Not far away a private farm road climbs from the blind road straight up a shallow grade and, near the top, make a peculiar bend is a small copse of trees, and locals claim that graves lie in their shade.

Years ago those graves were visibly sunken, and therefore easier for visitors to find; today they're harder to detect, and most be pointed out As long as anyone in the area can remember, the site has been protected by neighbors. Maybe this spot is indeed the final resting place of Riley Luther Scherer and Captain W. D. Denny. For all who still remember those two Texans who died at a crossroads so far from home, there is perhaps some comfort to be found that they rest in peace in a shady grove.[8]

8. William A. Palmer Jr., "Good Bye Dear Sister," The Unhappy War Of A Texan Who Was The Last To Enlist And The First To Fall," in *Civil War Times* (October 2016), 56-59.

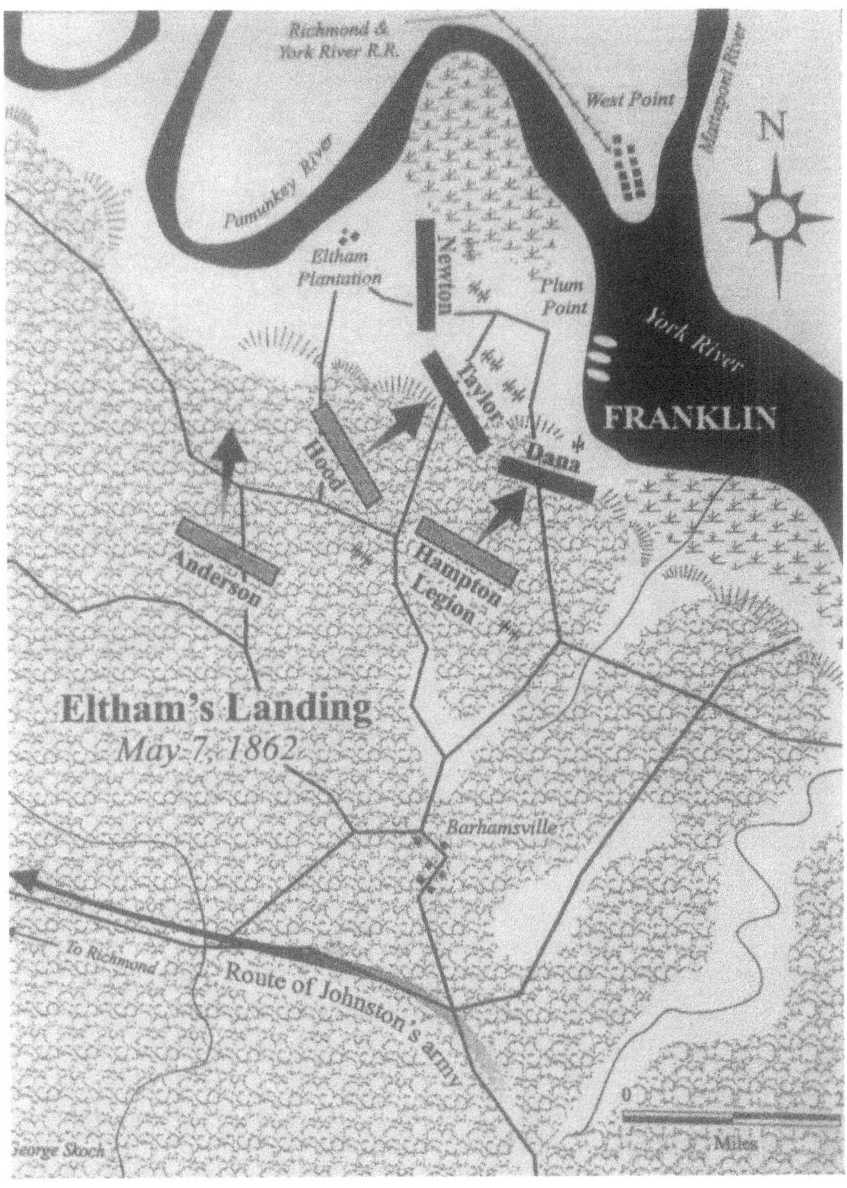

Map of the Battle of Eltham's Landing, May 7. 1862.

(George Skoch)

Flag of the 18th Georgia Infantry at Eltham's Landing.

(JackMelton.com Collection of Beverly DuBose III)

Chapter Four

18th Georgia

The 18th Georgia Infantry Regiment was composed of ten companies, mostly from central counties in Georgia.

List of Companies and First Commanding Officers:

Company A: (Acworth Infantry) Cobb County – J. L. Lemon
Company B: (Newton Rifles) Newton County – J. A. Stuart
Company C: (Jackson County Volunteers) Jackson County – D. L. Jarrett
Company D: (Davis Invincibles) Dougherty County – S. D. Irvin
Company E: (Stephens Infantry) Gordon County – E. L. Starr
Company F: (Davis Guards) Bartow County – J. C. Roper
Company G: (Lewis Volunteers) Bartow County – J. C. Maddox
Company H: (Rowland Highlanders) Bartow County – F. M. Ford
Company I: (Dooly Light Infantry) Dooly County – R. T. Coley
Company K: (Rowland Infantry) – Bartow County – J. A. Crawford

The 18th Georgia Infantry Regiment was organized at Camp Brown, Cobb County, Georgia on April 22, 1861 under a special act of the Georgia legislature and was originally designated First Regiment, Fourth Brigade, State Troops Commanded by Colonel William T. Wofford, and Lieutenant Colonel Solon Z. Ruff.[1] After drilling at

1. Folsom, J., *Heroes and Martyrs of Georgia. Georgia's record in the revolution of 1861*, Macon, GA: Burke, Boykin and Company 1864, 13. Lieutenant Colonel Solon Zachariah "S. Z." Ruff was born in Smyrna, Georgia in 1837. On April 18, 1861, he was appointed lieutenant colonel of the 18th Georgia Infantry, after the promotion of its previous commanding officer William T. Worford to brigadier general. He led his regiment in a successful charge against Union fortifications at the Battle of Gaines' Mill on June 27, 1862. He continued to lead the 18th Georgia when the regiment was detached from Hood's Texas Brigade after the Battle of Antietam. He was killed on the assault of Fort Saunders at the Battle of Knoxville on November 29, 1863, and is buried at the Gibson Family Cemetery near Knoxville, Tennessee. Owen, Joe, McBride, Philip and Allport, Joe, *Texans at Antietam:*

Col. William T. Wofford, Commanding Officer of the 18th Georgia Infantry.

(The House Divided Project at Dickinson College John Osborne)

Lt. Col. S. Z. Ruff 18th Georgia Infantry.

(Lucile Woods Hardwick Lind)

Camp McDonald, Georgia, the regiment was ordered to the Confederate capital Richmond, Virginia, arriving on August 8, 1861. Upon arriving, the regiments primary duty was guarding Union

A Terrible Clash of Arms: September 16-17, 1862. Charleston, SC: Fonthill Media Limited, 2017, 168.

prisoners captured at the battle of First Manassas fought on July 21, 1861. On November 20th, 1861 the regiment was assigned as the fourth regiment to the Texas Brigade, thus bringing the brigade to full strength under commanding General Louis T. Wigfall.[2]

Captain James Lyle Lemon, 18th Georgia Infantry.

(Mark Lemon)

Captain James Lile Lemon was a soldier in the "Acworth Infantry" militia in Acworth, Georgia. He joined the 18th Georgia Infantry Regiment as a 2nd Lieutenant and was promoted to 1st Lieutenant on June 5, 1862. He would fight in the major battles of Gaines' Mill, Second Bull Run (Manassas), Antietam (Sharpsburg), Fredericksburg, Chancellorsville, Gettysburg and Knoxville.

He was promoted to Captain on March 5, 1863, and was captured during the battle of Knoxville on November 29, 1863 and was first sent to POW Camp Chase, Ohio.8 Captain Lemon was then transferred to the POW Camp at Fort Delaware and arrived at the camp on March 27, 1864. Afterwards, he was sent to various POW Camps throughout the South, and would become a member of the "Immortal 600."

Captain James Lile Lemon
18th Georgia Infantry

The Yanks had landed and were bringing many men and horses and equipment ashore with every hour in preparation for their great campaign. This became known to history as the "Peninsula" campaign and had as its objective the capture of our capitol city, Richmond. We were at this time under the overall command of Gen'l Joseph Johnston, who was known then and later, for being a defensive minded general and he was soon to display his style. On the 4th of May 1862

2. Simpson, *A Compendium*, 323.

we were ordered to fall back from our lines, as those of the enemy were greatly strengthened and they threatened attack from both front (by land) and flank (by boat). Our brigade was assigned as the rear guard of the Army and held our position until the last of our forces had deported. We then marched all night at great speed. At length we arrived at Barhamsville, where we were to meet a force of the Yanks which had moved past our left flank by boat and were landing almost in our rear. Here, at Eltham's Landing, our Brigade saw its first real action. While our Regt was held in reserve, the remainder of the Brigade pitched into the enemy and drove him back, with great loss, over a mile to his boats upon the river. Our boys at once flew into a rage and not being actively used and claimed Hood was "playing favorites" with the Texans. I understood Col. Wofford made discreet inquiries and was told it was done due to our not yet being adequately armed and we will soon be supplied with better guns. The Colonel was assured we would soon get our chance, and thus we soon did, and with interest. Our Regt's part, such as it was, in the fight, was to capture about 40 Yankees as they fled. After the fight at Eltham's Landing, we again fell back as the Army's rear guard and arrived at our camp in Richmond about May 17th.[3]

Private Milton Barrett
18th Georgia Infantry

Kent County Virginia
May the 13th 1862

Dear Brother and sister hit is with pleasure that I take my pen in hand to let you know that I am well at present hoping these few lines may find you all injoying the like blesing. i have just received yours dated May the I and was glad to hear from you.

i was at Yorktown when i wrote last. sience then we have bin a marching a litle and a fighting a litle. we helt our fortification til the fourth of May having everything move to our rear. We vacated the place leaving nothing behind. our brigade was the last to leave. we layed in a line of battle two days and nights. hit taken that long for all

3. Lemon, Mark, *Feed Them The Steel! Being, the Wartime Recollections of Capt. James Lile Lemon, Co A, 18th Georgia Infantry CSA*, Self-published, 2013, 20.

the troops to get off. at 9 o'clock on Sunday moring we took up the line of march covering the retreat. we had not got many miles tel we found that a bout 600 yankee calvery was in per suit of us. we form a line of battle and our cavlery soon repulce them with thout any loss. by two oclock we got to Williamsburg all right learning that the Yankees was going in full speed up the river aiming to out flank us at west point. our brigade was put in front and hured to that place.

A grate many troops was stop a round Williamsburg. late on Sunday even the yankees advance on them and a considerable fight took place. hit lasted but few minits and sevrel fell on boath sides. i am not able to give the number but our troops drove them back and captured ther artilery.

we march nearly all night. one divishion stop at williamsburg. the rest move on towds west point. us in front landed with in six miles of that place by monday night.

the Yankees made a heavy actact on williamsburg at 11 on monday. tha was 3000 ingage on boath sides. the fight lasted tel 6 in the even. the yankees reinforce all the time and no reinforce got to our troops tel five in the even. our troops won the fight. drove the yankees back with grate slaughter. took 15 peces of artilary 1200 prisoners the loss was heavy on boath sides. we had 590 men kild and a large number wonded. the loss of the enemys was three times as large as ours. i am not able to give the corect account of the kild and wounded as; i have not he generals report of that fight. perhaps you have saw a correct account by now. hit was a hard and bloody battle.

We scouted a round with on the night of the sixth we learnt from our picets that a large number of Yankees had loaded men near West early on the moring of the 7. we march a few miles towards the rivor. gen Hood and staff was in front a looking wher to form his line of battle. a yankee picket fiard at him but mist him. the yankee was sot down on the spot. Two company of forth Texas was deploed as cromishers and was a throwing the yankees about right. the 18th georgia was in the rear of the artilery to defend hit. The other three Texas reg. advance thrue a strip o wood and was soon ingage with doble ther number. the first Texas command by Co. Rainy fought like wile cat and deserves grate honor. The fourth and fifth Texas fought bravley.

the 18th ga did not get to fiar a gun but took some prisnors. the first Texas had 12 kild, 19 wounded, me and 15 in our company was place out in a strip of woods to keep the sharp shooters off. the ball fell a round us and the bums burst among us but we escape unhurt. That was 1200 yankees kild and wounded and taken prisnors by our brigade.

Col Hampton took 200 prisnors and kild sevrel. we are expecting nother fight soon. we have fell back out of the reach of the gun boats on the Chickhomey rivor.

I remain your effecnate brother tel death.

Milton Barrett[4]

4. Heller III., J Roderick Heller (ed), and Ayers, Carolyn, (ed), *The Confederacy Is on Our Way up the Spout. Letters to South Carolina, 1861-1864*, Columbia, SC: University of South Carolina Press, 1998, 66-68. Private Milton Barrett enlisted in the 18th Georgia Infantry in Acton, Georgia on June 13, 1861, and participated in all the battles that involved the 18th Georgia up until his capture at Front Royal, Virginia. He was first imprisoned at the Old Capital Prison in Washington D.C., and shortly afterwards was transferred to the Elmira Prison in Elmira, NY. He died of smallpox on February 12, 1865 and is buried in the prison cemetery.

CHAPTER FIVE

Generals Correspondence

Gustavus W. Smith entered the Confederate service as a major general commissioned by the Confederate congress on September 19, 1861 He commanded a wing of the Army of Northern Virginia during the Peninsula Campaign. For a few hours, he was in command of the Army of Northern Virginia after the wounded of General Joseph E. Johnston at the battle of Seven Pines. Smith also served as Secretary of War ad interim in November 1862.

Maj. Gen. Gustavus W. Smith.

(Library of Congress)

Major General Gustavus W. Smith

Report of Maj. Gen. Gustavus W. Smith, C. S. Army, commanding Reserve.

Headquarters Reserve,
Cross-Roads, Va., May 12, 1862

Major: I have the honor to transmit herewith a report just received from Brig. Gen. W. H. C. Whiting, command a division corps, with reports of commanders under him, describing an engagement with the enemy on the 7th instant, between Barhamsville and Eltham's Landing – the later place being nearly opposite West Point on the south side of the Pamunkey River.

At Barhamsville the line by which my command was retiring formed on angle, at which the trains were exposed by the enemy. Owing to difficulties in getting artillery and baggage over the roads, I had been directed by General Johnston to halt there until the troops in the rear could be brought up. During this halt a fleet of transports, protected by gunboats, appeared at the head of York River, and from them the enemy commenced landing troops.

After examining their operations I elected a position for the division of General Whiting, and directed him to prevent the enemy from advancing upon Barhamsville until all the trains had passed.

The nature of the ground at the point chosen by the enemy for their landing and the protection afforded by their gunboats rendered an attack at that time and place not advisable, and I preferred to let them land and move out beyond the protecting range of their heavily-armed iron-clad vessels. The latter it was supposed they would attempt during the night or early morning.

The command proper of Major-General Magruder, then under Brig. Gen. David R. Jones, was placed at my control by General Johnston, and he directed the commands of Major-Generals Longstreet and Hill to be brought within close supporting distance.

On the morning of the 7th, after becoming satisfied that the enemy did not intend to advance in force from under the protection of the gunboats, I directed General Whiting to drive their skirmishers from the dense woods and endeavor to get position in the open ground between the woods and the river, from which he could reach their place of landing and their transports with his artillery fire.

After quite a sharp contest the enemy were driven back through the woods for a mile or more, when it was found that from the positions attained the range was too great for our fire to reach the transports, and that the troops and material already landed were completely covered from view by a bluff bank near the edge of the river. I then directed the troops to be withdrawn out of reach of the fire of the gunboats and to resume their position nearer Barhamsville. The enemy remained close under cover, protected by their gunboats.

Referring to the reports of the several commanders for details, it is only necessary for me to state that the Texas Brigade under command of Brig. Gen. John B. Hood, supported on the right by the Hampton Legion and the Nineteenth Georgia Regiment of Colonel Hampton's brigade, were selected and ordered forward by General Whiting, to drive the enemy from the woods then occupied in front of their landing. Later in the day the Tennessee Brigade, commanded by Brigadier-General Anderson, was placed in position to support and cover the left flank of the Texans.

All the troops engaged showed the finest spirit, was under perfect control, and behaved admirably. The brunt of the contest was borne by the Texans, and to them is due the largest share of the honors of the day at Eltham.

The Texas Brigade lost 8 killed and 28 wounded. In the other portions of the command there were 12 wounded and none killed.

The loss sustained by the enemy is not accurately known, but it was much greater than ours. The number of prisoners taken and sent to Richmond was 46. That night we continued our march without interruption to New Kent Court-House.

The affair at Eltham forms one of the most interesting incidents of the march of my command in retiring from Yorktown out of the Peninsula. The route is nearly parallel to the deep, navigable river, filled with vessels of war, gunboats, and transports of the enemy. Along this river are many most favorable landings, and good lateral roads leading from the river, intersecting our line of march at almost every mile, and at points varying in distance between 1 and 3 miles form the river. This delicate movement has been successfully accomplished.

The comfort and quiet with which the march of the troops has been conducted on this line is largely due to the admirable dispositions and watchfulness of the cavalry under Brig. Gen. J. E. B. Stuart, supported from day to day by brigades detailed for this purpose.

All of my staff officers have been untiring in their efforts to assist me in conducting this movement.

The whole command is in the finest spirits and in excellent condition, anxious to meet the invaders at any odds.

Respectfully, your obedient servant,

G. W. Smith,
Major-General Commanding.[1]

---

1. *The War of the Rebellion: A Compilation of the Official Records of the Union and Confederate Armies*, 128 Vols. (Washington DC, 1880-1901), Series 1, Vol. 11, pt. 1, 626-628. Here after cited as *OR*. All references are to Series 1 unless otherwise noted.

Brig. Gen. W. H. C. Whiting.

(Library of Congress)

At the outbreak of the war, William H.C. Whiting entered Confederate service as an engineer with the rank of major. He soon joined General Joseph E. Johnston's Army of the Shenandoah as chief engineer. Commanding a division, he fought in the battle of Seven Pines, and the Seven Days' battles. After the battle of Malvern Hill, he was ordered to Wilmington, North Carolina, where he developed Fort Fisher into the strongest fortress in the Confederacy. He was promoted to major general on April 22, 1863. For a brief time in the summer of 1864, he was assigned to Petersburg, Virginia. He failed to get his command into action at Port Walthall Junction, which earned him the accusation of being under the influence of whiskey or narcotics. For the rest of the war, he was assigned to North Carolina. He was severely wounded at Fort Fisher during a prolonged naval bombardment and land assault on January 15, 1865, and became a prisoner of war. He was sent to Fort Columbus on the New York Harbor where he died of his wounds on March 10, 1865.[2]

Brigadier General W. H. C. Whiting
Commanding General First Division

Report of Brig. Gen. William H. C. Whiting, C. S. Army, commanding First Division.

2. Warner, *Generals in Gray*, 334-335.

Hdqrs. Defense of Cape Fear, N. C.
November 14, 1862.

Major: Having learned that no record had been received at the War Department of the battle of Eltham's Landing, fought on May 7 last, by my division then forming part of the Reserve Corps, commanding by Maj. Gen. G. W. Smith, I beg leave to supply a copy. It is defective in wanting the report of the gallant Col. (now Brig. Gen.) Wade Hampton, which was with the original, together with the names of the killed and wounded. Otherwise it is word for word from the notes on which my original report was written.

We have since learned that the enemy's loses were very heavy in that battle. It was very creditable to the officers and men, and produced important results on the enemy's movements. It is due to the division that the record of one of the most brilliant of its many battles should be supplied on the files of the War Office. I have, therefore, to request that you will ask the major-general commanding the department to have this forwarded. If I am not mistaken, he himself sent in the originals with a report of his own.

Very Respectfully,

W.H.C. Whiting,
Brigadier-General, late Commanding First Division.

Maj. Samuel W. Melton,
Assistant Adjutant-General, Richmond, Va.

Headquarters First Division, Reserve Corps,
Camp, near New Kent Court-House, Va., May 8, 1862.

Major: The following are matters of interest connected with the engagement yesterday in the vicinity of Eltham's Landing:

On the 6th instant, at 12 m., I received intelligence that a portion of the enemy's fleet had anchored off of West Point and was preparing to land troops. Observing the boats carefully, I thought they might contain from twelve to sixteen regiments. The landing commenced very soon after they anchored, and light troops were thrown out to scour the woods; these were not interfered with. The major-general commanding the Reserve Corps rode out and personally examined the ground and approaches and observed the fleet. Upon that he directed a change of position of the First Division, which was accordingly made, the troops bivouacking in order of battle. During the night I ascertained that artillery was being landed, perhaps one battery and several regiments supposed to be one brigade.

Early on the 7th, the major-general directed me to attack the enemy, who was extending his line of pickets, and drive his advance back to the cover of his gunboats, to prevent any interference with the march of the main column. Accordingly, I directed Brigadier-General Hood, commanding the Texas Brigade, to advance on the Brick House and Barhamsville road and attack, while Colonel Hampton, commanding Second Brigade, should detach the Legion infantry and the Nineteenth Georgia to skirmish on our right, the Third Brigade being in reserve. They had hardly entered the timber when the fire was opened. The woods were very dense and extensive. From the moment of entering, the enemy, though several times reinforced, were steadily driven back by those brave troops. Two attempts were made to flank us in force – one on our left repulsed with great vigor by the First Texas, directed in person by General Hood, and one the right, beaten back by Colonel Hampton, the whole supported by General Anderson on the left and Third Brigade on the right – had driven the enemy fairly before it for over 1 ½ miles through a very dense forest, in which it was impossible to see over 30 or 40 yards. The coherence, discipline, and bravery of the troops were conspicuous. The fire of the enemy was heavy, but very high, which accounts for our small loss. That of my troops were deliberate and reserved, for though engaged for four hours, they did not expend over 7 or 8 rounds.

At 12 m. the enemy were driven under the cover of their gunboats, which opened at random and on the timber, but with no effect. Large numbers of their dead and wounded were left on the ground over which they were driven. While in this position one more effort was made by the enemy on our right, but speedily repulsed. I then ordered up Major Lee, with two rifled pieces, and Captain Reilly, with two Parrott (Manassas) guns, to occupy a bluff on the river and attempt to reach their transports. The battery was supported by the Sixth North Carolina, Colonel Pender, which had been posted all the morning in advance on our extreme right. The battery opened, but the range was too great, and the battery was withdrawn, the density of the woods making it useless on land. A gunboat got into position against the bluff just afterward, and got its range with great accuracy and rapidity, firing exceedingly well. The only effect, however, was the wounding of 2 of the Sixth North Carolina – not dangerously, however. Our ambulance train was then ordered up and the dead and wounded cared for. The enemy left many of his upon the field. The most painful cases of the latter were taken up by our surgeons and carried in to be attended to.

At this time (between 1 and 2 p.m.), the object designated having been accomplished, I ordered Brigadier-General Hood and Colonel Hampton to withdraw the line and take position in line of battle, to cover the march of the force from Barhamsville. This was done leisurely. Prisoners taken in action reported General Franklin to be in command of the enemy. Brig. Gen. John Newton of Virginia commanded the brigade at first opposed to us. I gather also, but am not sure, that one brigade of the enemy was under command of Brig. Gen. Philip Kearny.

To the conspicuous gallantry of Major-General Hood and Col. Wade Hampton, with the small number of brave troops immediately under their command, the credit of the engagement is due, and the result, which it was designated to effect, was that the large baggage ordinance, and artillery train was quietly and successfully moved, in perfect order, within three-quarters of a mile of 20,000 of the enemy upon our flank.

My thanks are due to all of my staff for their service during the day; to Maj. J. H. Hill, assistant adjutant-general, and Lieutenant Strong, aide-de-camp, in disposing the reserves; especially to Colonel Upson, of Texas, and Captain Vanderhorst, of South Carolina, aides, and Captain Frobel, of the artillery, who accomplished the advanced troops wherever they found the enemy, and to the troopers of the Hampton Legion, under Sergeant Beattie, attached to headquarters, all of whom brought me important information and did their duty well.

I take occasion to make my acknowledgements to Brigadier-General Anderson, of Tennessee, who, arriving on the field at a critical moment to the support of General Hood, and placing two of his regiments in the fire of the enemy, courteously waived the command, although senior to us all. I am informed that a few of his soldiers were wounded.

I transmit herewith the reports of Brigadier General Hood and Col. Wade Hampton, by which you will see that our loss is very small – 8 killed, 32 wounded, and none missing. We took 46 prisoners.

Very respectfully,

W. H. C. Whiting,
Brigadier-General, Commanding First Division of Reserves.
Maj. Jasper S. Whiting, Assistant Adjutant-General.

General W. H. C. Whiting

WILMINGTON JOURNAL
JUNE 5, 1862

Gunboats.

The following communication from Brigadier General Whiting contains some interesting facts concerning gunboats. It shows that their capabilities are very much magnified, and that they are not the all-powerful and irresistible things many prominent persons even at the South have regarded them to be. The lesson taught at the obstruction of the James river by our three gun battery, has helped to dispel the delusion that had prevailed on the subject. If the powerful vessels then essaying to reach the capital [one of whose officers informed an old acquaintance onboard the boat that carried the Yankee prisoners down the river, that he would dine in Richmond the day after the bombardment commenced!] were unable to pass while the obstruction was not complete, and when it was defended by only three guns in battery, what chance do those boats stand now? It is clear that the enemy can never reach Richmond by the river. He has had his gunboat conceit taken out of him to that extent. It is said that Lincoln was at Fortress Monroe and ordered the boats to proceed at once and take Richmond in advance of McClellan. But even Lincoln could not force his terrible boats by the small battery at Drewry's Bluff – a barrier which he has "hearn on" before this. We invite attention to General Whiting's letter:

Headquarters 1st Division Reserve
Near Richmond, May 23, 1862
To the Editor of the Dispatch:

Sir: - The panic cause by the enemy's gunboats is not, perhaps unreasonable, considering the manner in which their deeds have been trumpeted and exaggerated; but it is very absurd and needless. I will state a few facts: - Horizontal, or rather direct shell firing, as from Dahlgren or rifled places, is destructive only to shipping or buildings. It has very little effect upon batteries, either with or without parapets, or upon troops deployed. This is due, I think, partly to the difficulty of causing the shell , even if well directed, to explode exactly at the object, but chiefly to the great velocity of the missile, which carries the fragments after explosion beyond the point of bursting. Mortar firing, or vertical fire, as it is called, is, if well managed, much more destructive and dangerous, in that velocity is small and the explosive force has full play. I have never known a shell from the gunboats, exploding over an open battery or group or line of men, to do any harm.

Last winter a portion of my command, under Capt. Frobel, was exposed to frequent day and night shell firing of 11 and 9-inch Dahlgren and rifled shells. The camp was shelled repeatedly, and fire started in it once by some camphene contrivance in the missile. We had but one casualty.

During the same time, the enemy's gunboats fired between one hundred and two hundred heavy shell at a single picket station on the Potomac, without either hurting a man or driving off the guard.

The gallant Major Walker, of the artillery, unlimbered his battery in the open field, and for several hours engaged the Pawnee, Live Yankee, Anacostia, and a tug, at long range, without a single casualty, compelling the boats to retire. They were commanded by a brave officer of the U. S. N. – Capt. Rowan – who is reported to have lost an arm in the fight.

The battery at Acquia Creek was shelled for three days last summer by the Potomac fleet, one vessel alone expending several hundred shells with no other effect than a scratch on one man's hand and the killing of a trooper's horse in the rear of the battery.

Major Walker was over twelve times last winter and fall under fire of9-inchDahlgrensfromgunboatwithhisbattery,withoutacasualty.

Lieut. Col. Stephen Lee, of the Hampton Legion Artillery, now commanding the artillery of this division, twice engaged them – field battery against 9-inch shells – without a casualty.

When the Evansport batteries were opened on the Seminole and Pocahontas, even in the very imperfect condition of these batteries at that time, the enemy's fire inflicted no damage whatsoever.

At Yorktown, although the gunboats fired a great deal into the town and at Gloucester Point, both of which were full of tents, hospitals, storehouses, wagons, and troops, but one man was killed at Yorktown, and nobody hurt at Gloucester Point.

The other day, at the engagement between a part of my command and the enemy at Eltham's Landing, near West Point, one of the enemy's gunboats opened upon a position from which a few moments before Lieut, Col. Lee and Captain Reilly had in vain endeavored with four pieces to reach the transports; and though the boat got range with great accuracy, and burst her shells directly over the heads and within a few feet of the 6th North Carolina regiment, commanded my Col. Pender, but one man was hurt; and although their gunboats shelled the woods in which Hood's brigade and the Legion were drawn up, after the enemy were driven out, not a man was touched by them.

I can vouch for most of the above circumstances from personal knowledge and observation. Officers of my command attest all I have not seen myself.

At Port Royal and Hatteras the result was due partly to overwhelming force, and partly to inexperience on the part of the gunners. It is said – and I have no doubt of it – that many guns were dismounted by the recoil, owing to the eccentrics having been carelessly left in gear when the pieces were fired, while others were spiked and rendered useless by the priming wires being left in the vent, while the charge was rammed. Our firing was also bad. Yet, at both these places the loss of the garrison was trifling, when compared to the weight and length of the enemy's fire.

At Donelson, by Commodore Foote's own showing, the gunboats were fairly whipped and withdrew. – Gunboats had nothing to do with the taking of Fort Pulaski. At Island N. 10, they failed. Forts Jackson and St. Phillips surrendered from the treachers and mutiny, and that alone. Fort Macon was taken by land batteries.

Let our men stand to their guns and fight them as the officers and men of our Navy, did the other day at Drewry's Bluff, and there are few rumors in this country in which the Yankee gunboats will venture far – certainly not the James, the Cape Fear, or the Southern inlets below Charleston.

I hope the above statement will not cool somewhat of the excitement produced by the name and approach of gunboats. As far as our side is concerned, there can be no doubt as to the accuracy of the above statements. – What effect we may have produced on the enemy, there is no means of ascertaining. In time, however, from the false and contradictory statements they are compelled to make, the truth is gradually sifted. Thank God, our leaders are not obliged to lie either to keep our courage up or to steady our Government or people.

Very Respectfully,
W. H. C. Whiting,
Brig. Gen. Comm'g 1st Div. Reserve,
Army of the Potomac.

Brig. Gen. John Bell Hood

(John Bell Hood Papers Memorial Hall Museum New Orleans, LA)

Graduated from West Point in 1853, and served in California, and in the famous 2nd U. S. Cavalry Regiment in Texas under Colonel Albert Sidney Johnston and Lt. Colonel Robert E. Lee. He immediately resigned his commission in the U. S. Army after the firing upon Ft. Sumter and cast lot with the Confederate Army. At the battle of Gaines' Mill, heled his old regiment, the 4th Texas Infantry on a successful charge against Union fortifications. As division commander under General Longstreet, he distinguished himself in the battles of 2nd Bull Run (2nd Manassas,) Antietam (Sharpsburg,) and Fredericksburg. During the battle of Gettysburg on July 2, 1863, Major General Hood was severely wounded in the left arm by an exploding artillery shell above his head while leading his brigade in the assault of the Devil's Den and Little Round Top.

At the battle of Chickamauga on September 20, 1863, he was again severely wounded, this time, in the right leg by a minnie ball, and was reported dead on the field. However, the doctors were able to save him with the cost of amputating his leg.

He was appointed lieutenant general on February 1, 1864, and full general with temporary rank on July 18, 1864. He led his corps that was defeated at the battle of Franklin on, November 30, 1864, and six of his generals would be killed in action on the same day. He reverted back to his rank of Lieutenant General and surrendered himself in Natchez, Mississippi where he was paroled on May 31, 1865. After the war, Hood and his family resided in New Orleans, where on August 30, 1879 he died from yellow fever along with his wife and one of their children.[3]

3. Warner, *Generals in Gray*, 143.

Brigadier General John B. Hood
Commanding General Texas Brigade

Report of Brig. Gen. John B. Hood, C. S. Army, commanding First Brigade.

Headquarters Texas Brigade
Near Barhamsville, Va., May 7, 1862.

Sir: I have the honor to report that at 7 o'clock this morning, agreeably to your instructions, Col. J. J. Archer, with his regiment, Fifth Texas, of this brigade, proceeded on the blind road leading to Eltham's Landing, on the Pamunkey River, to reconnoiter and drive in the skirmishers of the enemy. He soon met them and drove them steadily in front of him.

I at once proceeded with the remainder of the brigade, Col. John Marshall's (Fourth Texas) regiment, Col. A. T. Rainey's (First Texas) regiment, Col. W. T. Wofford's (Eighteenth Georgia) regiment, Lieut. Col S. Z. Ruff commanding, and the battery of Capt. W. L. Balthis, on the road leading from New Kent Court-House to this landing.

On arriving within some 20 paces of our cavalry pickets the enemy suddenly appeared deployed as skirmishers, and immediately opened their fire. The Fourth Texas Regiment was in front and their arms unloaded, as I had not thought it necessary to load so long as I was within our line of pickets. They, however, soon loaded under the enemy's fire and drove them back into the timber. Leaving at this point the battery and the Eighteenth Georgia Regiment, I threw forward the Fourth Texas as skirmishers, supported by the First Texas Regiment, driving the enemy through a dense forest with considerable loss.

The enemy during this time were re-enforced and placed in position to receive me. The First Texas Regiment was accordingly attacked with a terrible fire on its flank by two regiments. I immediately threw one wing of this regiment back and the other forward, which caused some little confusion, which being soon rectified, they, with Captains Porter's and Martin's companies of the Fourth Texas Regiment, and a platoon of Captain Carter's company, of the same regiment, charged gallantly forward, driving the enemy in utter confusion in front of them. Immediately after this Colonel Archer came up with his regiment and took position on the right line of battle.

Having driven the enemy a distance of 1 ½ miles, through a most difficult forest, forcing him under the protection of his gunboats, I then at 2.30 p. m., according to instructions, gathered up the killed and wounded and returned in perfect order to the bivouac I left in the morning.

I would respectfully state that Colonel Hampton, with about 400 of his Legion, forced the enemy on the right to return to the protection of their gunboats and that General Anderson arrived about noon with two regiments and held securely my left flank.

I captured some 40 prisoners and secured 84 stand of arms. The density of the forest and the large area over which the engagement extended prevented my securing more of the latter without permitting my men to straggle.

The force against me was one brigade in the beginning, and I am of the opinion that it was considerably re-enforced.

The loss of the enemy in killed and wounded I think was heavy, and I think, from personal observation, that it could not have been less than 300, although it was impossible to approximate to the exact number in consequence of the facts already referred to.

Capt. W. H. Sellers, assistant adjutant-general, and Lieut. D. L. Sublett, aide-de-camp, rendered me most efficient services in bringing forward the troops and transmitting orders.

The conduct of officers and men, one and all, was beyond all praise.

My attention was particularly called to the great gallantry of Captain Decatur, of the First Texas, who fell under the heavy fire upon the flank of his regiment.

I am, sir, very respectfully, your obedient servant,

J. B. Hood,
Brigadier-General, Commanding Texas Brigade.[4]
Maj. James H. Hill, Assistant Adjutant-General.

Brigadier General John Bell Hood
Commanding General Texas Brigade

I had been stationed a few weeks in the vicinity of Fredericksburg, when orders were received to march to Yorktown, at which place we arrived a few days prior to the 17th of April, the date of General Johnston's assumption of the command of all the forces on the Peninsula. I was here placed in reserve with my brigade, which

4. *OR*,11, pt 1. 631-632.

consisted of the First, Fourth, Fifth Texas, and Eighteenth Georgia Regiments, and continued the system of instruction and training already indicated. I had so effectually aroused the pride of this splendid body of men, as to entertain little fear in regard to their action on the field of battle.

The 3d of May, "on information that the Federal batteries would be ready for service in a day or two, the Commanding General ordered the Army to retreat. Accordingly, I marched with my brigade, which formed part of Major General G. W. Smith's Division, upon the Yorktown road, in the direction of Williamsburg. At daybreak of the 5th the retreat was continued from Williamsburg towards Richmond, through deep mud, and in a heavy rain. Whilst in bivouac opposite West Point, General Whiting informed me that a large body of the enemy had disembarked at Eltham's Landing; that our cavalry was on picket upon the high ground overlooking the valley of York river, and instructed me to move my brigade in that direction, and drive the enemy back if he attempted to advance from under cover of his gunboats. Pursuant to imperative orders, the men had not been allowed to march with loaded arms during the retreat On the 7th. at the head of my command, I proceeded in the direction of Eltham's, with the intention to halt and load the muskets upon our arrival at the cavalry outpost. I soon reached the rear of a small cabin upon the crest of the hill, where I found one of our cavalry- men half asleep. The head of the column, marching by the right flank, with the Fourth Texas in the front, was not more than twenty or thirty paces in my rear, when, simultaneously With my arrival at the station of this cavalry picket, a skirmish line, supported by a large body of the enemy, met me face to face. The slope from the cabin toward the York river was abrupt, and, consequently, I did not discover the Federals till we were almost close enough to shake hands. I leaped from my horse, ran to the head of my column, then about fifteen paces in rear, gave the command, forward into line, and ordered the men to load. The Federals immediately opened fire, but halted as they perceived our long line in rear. Meanwhile, a corporal of the enemy drew down his musket upon me as I stood in front of my line. John Deal, a private in Company "A," Fourth Texas Regiment, and who now resides in Gonzales, Texas, had fortunately, in this instance, but contrary to orders, charged his rifle before leaving camp ; he instantly killed the corporal, who fell within a few feet of me. At the time I ordered the leading regiment to change front forward on the first company, I also sent directions to the troops in rear to follow up the movement and load their arms, which was promptly executed. The brigade then gallantly advanced, and drove the Federals, within the space of about two hours, a distance of one mile and a half to the cover of their gunboats. When we struck their main line quite a spirited engagement took place, which, however, proved to be only a temporary stand before attaining the immediate shelter of their vessels of war. Hampton's brigade, near the close of the action, came to our support, and performed efficient service on the right.

Our loss was slight, whereas that of the enemy was quite severe. General Johnston states in his Narrative that if North- ern publications of that period are to be relied upon, it was ten times greater than our own. The Commanding General of the Army, though correct in his assertion that the security of his march required the dislodgement of the enemy from its position south of Eltham's Landing, is in error in regard to the troops who bore the brunt of the combat, as will be seen by the following extract from the official report of Major General G. W. Smith, who at that time commanded the division:

Referring to the reports of the several commanders for details, it is only necessary for me to state that the Texas brigade, under command of Brigadier General John B. Hood, supported on the right by the Hampton Legion and the Nineteenth Georgia Regiment, of Colonel Hampton's brigade, were selected, and ordered forward by General Whiting, to drive the enemy from the woods then occupied in front of their landing. Late in the day the Tennessee brigade, commanded by Brigadier General Anderson, was placed in position to support and cover the left flank of the Texans. All the troops engaged showed the finest spirit, were under perfect control, and behaved admirably. The brunt of the contest was borne by the Texans, and to them is due the largest share of the honors of the day at Eltham. The Texas brigade lost eight killed and twenty- eight wounded; in the other portions of the command there were twelve wounded and none killed." This affair, which brought the brigade so suddenly and unexpectedly under fire for the first time, served as a happy introduction to the enemy.[5]

Driving the Enemy into the River.

FORT HOOD SENTINEL
JULY 2, 1992

At Eltham's Landing on the York River, May 7, 1862, a force of federals estimated at three to five thousand were disembarking from gunboats. Hood's orders were to "feel the enemy gently and fall back,

5. Hood. J. B., *Advance and Retreat: Personal Experiences in the United States and Confederate Armies*, New Orleans, LA: Beauregard, 1880, 22-23.

Colonel Wade Hampton.

(Library of Congress)

avoiding an engagement, and draw them away from the protection of those gunboats..."

Hood found the federals already away from the gunboats and attacked - driving them back a mile and one half until they were again under the protection of the boats. His Texans killed or wounded 300 and captured 126. The Texans lost was 37 killed or wounded.

General Joseph E Johnston, commanding the Confederate forces, said, "Gen. Hood you have given an illustration of the Texas idea of feeling the enemy gently and falling back. What would you Texans have done sir, if I had ordered you to charge and drive back the enemy?"

Hood replied, "I supposed, general, they would have driven them into the river and tried to swim out and capture the gunboats."

Wade Hampton, III served in both houses of South Carolina legislature from 1852 to 1861. In 1861 he was reputed to be the largest landowner in the South.

At the outbreak of the Civil War he organized the Hampton Legion and became its colonel. He equipped the Legion with his own expense, and took it to Virginia in time to participate in the battle of First Manassas, where he was wounded. He was appointed brigadier general on May 23, 1862. In July he assumed command of a brigade in J. E. B. Stuart's Cavalry Corps and participated in most of Stuart's operations in 1862. He was severely wounded at the battle of Gettysburg; afterwards he was promoted to major general on August 3, 1863.

After the death of General Stuart, he assumed command of the Stuarts Cavalry Corps. He and his Corps performed outstandingly in keeping the Federal cavalry outside of Richmond and Petersburg until winter 1864. He was then ordered, with part of his corps, to General Joseph E. Johnston in the Carolinas. He was elected governor of South Carolina in 1876 and would later serve as a United States Senator from 1879 to 1891.

Colonel Wade Hampton
Hampton Legion/Second Brigade Commander

Report of Col. Wade Hampton, Hampton (S. C.) Legion, commanding Second Brigade.

Headquarters Second Brigade,
_______, ____ ___, 1862.

Major: In pursuance of orders from General Whiting to take the Legion into the wood adjacent to the Brick-House Ferry, on the morning of the 7th instant I ordered the infantry battalion under Lieutenant-Colonel Griffin, to proceed on the on the road from Barhamsville to York River and to attack the enemy, reported to be then in occupation of the woods. The Nineteenth Georgia Regiment, Lieutenant-Colonel Johnson commanding, was ordered to take position near the entrance of the road into the woods. The two flank companies of this regiment were detached to act as skirmishers, together with two from the Legion, the four being laced under the command of Major Conner, of the Legion. Major Lee was directed to hold four rifled pieces in readiness either to support the infantry or to attack the shipping of the enemy if this latter was found practicable. The skirmishers entered the wood, two companies on each side of the road, followed by the infantry of the Legion. As soon as we entered the woods General Hood's command opened fire on the enemy on our left, and soon afterward I fell in with the Fifth Texas Regiment, Colonel Archer commanding, which had got into the same road we were pursuing. Fearing a collision between my command and that of General Hood, should they be thrown together in the thick woods, I acceded to the request of Colonel Archer, and allowed him to proceed me until he had formed a junction with General Hood. After that the line was drawn at right angles to the road. General Hood being on the left and my command on the right of it. In this order we advanced, driving back the enemy, who seemed to be re-enforced constantly, to the very edge of the woods. As soon as their own men were clear of the woods the enemy opened fire on us from their gunboats and a field battery in our front. Anticipating an attempt to turn my right flank, I sent, by permission of General Whiting, for the Nineteenth Georgia, which came to me at double-quick and took position rapidly and steadily on my right, though the fire of the enemy was then very heavy. Soon after they had taken position an order came directing them to move to the support of Major Lee, who had opened fire on the vessels of the enemy.

General Hood had withdrawn his command by this time and before I could withdraw the Legion a strong attack was made on the line of Colonel Griffin, which resulted in wounding 4 of my men. My men returned the fire, and responding to my order to charge with a cheer, drove the enemy back.

Colonel Griffin was now left quite alone, and as soon as the wounded could be placed in the ambulances I withdrew the Legion from a position which had become very precarious.

I am happy to say that I lost in killed no men; 4 were wounded.

As I sent back various prisoners taken by the Texans, I am not quite sure of the number taken by my men. My officers of the Nineteenth Georgia report the number as 12.

I take great pleasure in saying that the conduct of officers and men met my entire approval. Colonel Griffin, in command of the Legion, handled them admirably, while Major Conner did the same with his four companies of skirmishers.

Colonel [Thomas C.] Johnson and Maj. A. J. Hutchins, of the Nineteenth Georgia, behaved as well as I could desire, while Major Lee again displayed the soldierly conduct for which he is conspicuous.

I have the honor to be, very respectfully, your obedient servant,
Wade Hampton,
Colonel, Commanding Second Brigade.

Maj. James H. Hill,
Assistant Adjutant-General.[6]

He entered the U. S. Army as captain of Company G, 16th New York Infantry. He was severely wounded leading his company at the battle of Eltham's Landing (West Point) on May 7, 1862. He participated in the battles of Cold Harbor and Petersburg as part of the Army of the James. In January 1865, he was part of the successful storming of Fort Fisher, and was awarded the Congressional Medal of Honor for being the first Federal soldier inside the ranks. After

6. *OR*, Vol. 11, pt. 1, 632-633.

receiving the Medal of Honor, he was quickly promoted to brigadier general in the regular army, and brevetted major general of volunteers. After the war he served in the House of Representatives.

THE DAILY EXPRESS
APRIL 15, 1906

Department of History

Address all letters Intended For This Department to J. B. Polley, Floresville, Texas.

Maj. Gen. Newton Martin Curtis

(Library of Congress)

Readers will doubtless remember the publication in the issues of the Express of March 11 last of an account of the battle of Eltham's Landing in Virginia, where in the Texas Brigade under Gen. John B. Hood played a prominent part. A copy of that account was forwarded to General Newton M. Curtis, who by the way, is known by the G. A. R. as the hero of Ft. Fisher and in response he wrote the following letter and furnished the following copies of articles pertinent to the subject. General Curtis has a book in press, wherein he related his experiences and conclusions during and concerning the War Between the States. It will be, beyond doubt, a valuable contribution to history:

20 Irving Place, N.Y., March 6, 1906. – Capt. J. B. Polley, Floresville, Tex.

- Dear Captain: I thank you for your letter of the 28th ultimo, which I found awaiting my return from Washington.

I do not recall the events of May 7, 1862, at West Point, as we call it and which I know is an erroneous designation, as West Point is on the left bank of the Pamunkey, while we landed on the right flank at Brick

House Point near Eltham's Landing, as you describe them in the interesting article you sent me, I enclose the chapter of my book which treats of the battle of West Point and I will be very glad if you will examine it and offer many, many suggestions of information about the cases of Mummery, Seabury and Cook. I give the authority for the information gained when in Texas in 1869, but I cannot give the names of the men as the notes taken at the time have been lost. They were men of good appearance and I advised their employment in Western Texas, and afterward learned they discharged the duties of the positions assumed satisfactory.

General Whiting reported that your loss was 8 killed and 32 wounded, (page 629 W. R. V. XI. Pt. D.) That you captured 46. We reported casualties as follows:

Killed, 7 officers and 41 men; wounded 6 officers and 104 men; missing 28. Eighty percent of your losses were sustained by the Thirty-first and Thirty-second New York Regiments, Companies E and G of the Sixteenth. We thought that our troops behaved well. But I am not inclined to discuss the question. As an American I am proud of the valor, fortitude, and devotion of the men of both armies. They did what each thought was right, and that, too, in a manly and heroic manner.

The incidents related about Mummery and Cook took place some distance from each other. The Texans I saw had no recollection of the Masonic experience of Corporal Cook. That it took place as related I have no doubt as Cook is a man of high character. The men of the Sixteenth in battle were on or near the left of our line, that is where the main fighting was done. The Thirty-first and Thirty-second were on our right. The Pistnam's have my manuscript, and I hope to have the work done in a few weeks. If I get anything from you I can use I will be glad to insert it, if not too late. I have a letter from a member of the Twentieth North Carolina, which was hotly engaged with the Sixteenth at Gaines' Mill, June 27, 1862, and insert it in connection with the description of that battle. True history is what we all want, and written, too, without reviving business. It will interest you to read Gen. John B. Gordon's reference to me in his "Reminiscences of the Civil War." I have been for peace since Appomattox, and have warm friends throughout the South.
Very truly yours,
N. M. Curtis

Chapter IX. W. R. Vol. XL. Pt. 1. Pages 614 to 633.[7]

7. *OR*, Vol. 11, pt 1, 614-633.

The Battle of West Point.

After landing at the head of the York River, the Sixteenth Regiment marched a short distance and stacked arms. After supper was over the members of Company F were engaged in general conversation, when Edwin R. Bishop, a light-hearted, fun provoking man, rose from the ground and interrupted the conversation by saying: "Boys, if I shall fall in the next battle, as I now believe I shall, I wish you would bury me under this tree, where I indicate by these lines." He then proceeded to mark with a spade the outlines of a grave. Immediately Corporal George J. Love, a very sedate man, rose and picking up the spade which Bishop used and said: "I would like you to dig my grave besides Bishop's but please dig it with more regularity than his crooked lines indicate." He then proceeded to draw a parallelogram, dropped the spade and said, "If I fall, dig my grave here and do it as we dig graves at home. I am the son of a sexton and have helped dig many, please follow the lines I make for you." He drew the lines of the coffin used in those days, under at the shoulders and tapering toward the head and foot. Conversation was resumed and no further attention was paid toward the incident.

At 3 o'clock the next morning Companies F and G were ordered out to the picket line, where, at 9 a.m., they met the advancing lines of Gen. J. B. Hood's brigade of Whiting's division. These companies could not stay the progress of the overwhelming force brought against them, but made a manful resistance until the artillery was brought up and ready for action; they were then ordered back with 17 per cent of their number killed – Bishop, Love and Ploof and their comrades, in paying them the martial honors due the gallant dead, gave to each a resting place he had selected on the night before the battle. Beside them were buried Mummery, Seabury and Waymouth of Company G.

Corporal James Cook of Company F, whose leg was broken by a musket ball, was left on the field during its temporary occupation by the enemy; a Confederate soldier took his watch, purse and Masonic ring. His call for hop brought to his side a Confederate Mason, who caused Cook's property to be restored to him, filled his canteen with water, made him as comfortable as possible, and on having said: "We are enemies in honorable warfare, but on the plane where your disabilities have placed you, the laws of humanity and charity shall prevail."

Of the members of Company G, Seabury was found alive, but lived only long enough to tell his comrades that the Confederates had been kind to him, and had done all they could to make him comfortable. Weymouth had evidently been killed in the act of loading his musket; Mummery's body was found in a pool of water with his throat cut.

Great indignation was felt by all, and General Newton, in his report of the battle, referred to his case and others of less savagery, in terms of severe condemnation. That Mummery's throat should have been cut, when his wounds were mortal was a mystery, which remained unsolved until, in February 1869, I visited Texas. On the steamer, crossing the Gulf of Mexico to Brazos de Santiago, I fell in with two Texans who were in Hood's Brigade, and in the battle of West Point. I questioned them about the battle, and asked them to recall any unusual circumstances connected with it.

"There was nothing unusual," the spokesman said, "We found out that the Yanks would fight, and were not to be driven with pop guns, as we were told when we joined Magruder's army at Yorktown."

The other added, "That this was the place where we cut the Yank's throat."

He went on to tell the action, of their occupying the ground which we held at the beginning of the engagement, and said: "One, who was severely wounded and unable stand, opened on the Confederates with a seven-shooter, every shot of which killed or wounded a man. It was thought that a wounded man, who thereafter, continued the fight on his own account, deserved to be summarily dealt with, so we cut his throat."

It had been learned, after Mummery's death, that he disobeyed orders in not turning in his pistol at Alexandria, and that he confided to a comrade his purpose never to be captured alive, but to inflict all the injury possible on the enemy. There are many cases reported where disabled men have continued to fight after the opposing forces occupied the ground, and in nearly all instances, they became the subjects of summary treatment; a case of this kind occurred late in the war with Spain when a Spanish officer shot Lieutenant Ord and was promptly dispatched by a volley from Ord's company.

I quote from letters, and from official reports of the action.

Headquarters Sixteenth New York,
Brick House Point, York River, Va., May 8, 1862.

General:

I have the honor to report the part taken by the regiment under my command, in the engagement yesterday.

About 9 o'clock a.m. yesterday I received orders from Brigadier General Slocum to report with five companies, (C, D, H and I) to General Franklin on the right of the line. Companies A, B, F, and K were on picket, A, B and K posted the night before, and F and G having reported to the general officer of the day, Colonel Bartlett, Twenty-seventh New York at 3 o'clock a.m., and had been sent to relieve a portion of the advance guard from the Twenty-seventh New York, at our center and left. While the battalion under my command was marching to the front, I was ordered by General Slocum to support Captain Platt's battery, which was advancing near me, and to report to General Newton. Captain Platt took a position beyond a small stream, which empties into the York River on our left, and on the right of the road which leads inland. I placed my battalion in column on the left flank of the battery and a little in rear, but received orders from yourself to move to the left of the road and within supporting distance, where I would be hidden from the enemy by the woods, in case he made his appearance. I subsequently received orders through an officer of your staff to recross the stream and take positions further to the rear. As the execution of this ordered was received through Captain Scofield of your staff for the infantry to cross the stream when I took position in column on the right of the road and on the left flank of Hexamer's battery, which had come and taken the position previously occupied by Colonel Platt. I remained on this position until about 5 o'clock p. m., two hours after the artillery fire ceased, and was at no time under fire; but Companies F and G were engaged early in the day as skirmishers, while on duty at the outposts, and met with the same losses. As these companies were at the time detached from the regiment I enclose the reports of the company commanders. I have every reason to believe that the companies behaved well, and only fell back when obliged to do so by greatly superior forces, from want of

Colonel Joseph Howland, Commanding Officer of the 16th New York Infantry

(Library of Congress)

support, and on account of the imminent danger of being outflanked and surrounded.

Companies A, B, and K, upon being relieved as pickets, returned to camp for food and then started to rejoin their regiment, but on the way were ordered by Colonel Bartlett, commanding General Slocum's brigade to support Captain Wilson's battery, First New York Artillery. They were not engaged and received orders to return to camp about 5 o'clock p.m.

I have the honor of enclosing a list of killed and wounded. The wounded were invariable robbed, and in nearly every case were stripped of their jackets.

I am, general, very respectfully,

Your obedient servant,

Joseph Howland
Colonel Sixteenth New York.

Washington D. C., July 31, 1904.
Dear General Curtis:

As I now recall the action of May 7, 1862, at West Point, Va., my company held the left of the line and was deployed as skirmishers with a mall part of it in support on reserve. On my right was your company formed in the same way as mine, and at the time of starting was under the command of Lieut. S. C. Vedder of your company. I am not at this time sure, whether we were advancing or had halted when the fight commenced with the enemy in much stronger force than ours. While my company was holding them in check in my immediate front, they, by strong force drove the left of Vedder's line back, advancing with your company, having joined it from another part of the field, and drove the enemy from my left. I making this movement, you received a severe wound in your left breast, but kept the field until we were ordered to retire, which order was given as soon as the artillery was put into position to open fire. Your prompt action in coming to my aid saved without a doubt my command from greater loss than it sustained, which was three killed and five wounded.

Corporal James Cook of my company was severely wounded in the early part of the engagement and fell into the hands of the enemy. Later in the day we recovered the bodies of our killed and brought off the wounded except Barnhart and Kelley of my company, who were carried to Richmond.

After the regiment went into camp on the 11th of May, General McClellan rode up to our regiment headquarters and requested Colonel Howland to send for the Captains of the two companies engaged at West Point that he might thank them in person for their good conduct in the engagement. When informed that Captain Gilmore was on picket and Captain Curtis on a hospital boat in the York River he asked Colonel Howard and gave me General McClellan's message and informed me that he had communicated the same to you by mail.

Yours cordially,

Joseph Howland

Brig. Gen. John C. Gilmore

(Library of Congress)

John C. Gilmore
Brig. Gen. U.S.A. Ret.

From "Advance and Retreat" by Lieutenant General J. B. Hood, C.S.A., page 21.

"Whilst in bivouac opposite, West Point, General Whiting informed me that a large body of the enemy had disembarked at Eltham's Landing, that our cavalry was upon the high ground overlooking the valley of the York River, and instructed me to move my brigade in that direction and drive the enemy back if he attempted to advance form under cover of his gunboats.

"Pursuant to imperative orders the men had not been allowed to march with loaded guns during the retreat. On the 7th, at the head of my command. I proceeded in the direction of Eltham's, with the intention to halt and load the muskets on our arrival at the cavalry outpost. I soon reached the rear of a small cabin upon the crest of a hill, where I found one of our cavalrymen half asleep. The head of the column marching by the right flank with the Fourth Texas in front, was not more than twenty or thirty paces on my rear, when, simultaneously with my arrival at the station of the cavalry picket, a skirmish line supported by a large body of the enemy, met me face to face. The slope from the cabin toward the York River was abreast and consequently I did not discover the Federals ill they were almost close enough to shake hands. I leaped from my horse, ran to the head of the column, then about fifteen paces in my rear, gave the command forward into line and ordered the men to load. The Federals immediately opened fire, but halted as they perceived our long line in the rear. Meanwhile a corporal of the enemy drew down his musket upon me as I stood in front of my line. John Deal, a private in Company A, Fourth Texas, had fortunately in this instance, but contrary to orders, charged his rifle before leaving camp. He instantly killed the Corporal, who fell within a few feet of me."[8]

8. Hood, *Advance and Retreat*, 21.

Hood's Texas Brigade Monument, State of Texas Capital Grounds.

(Diane Kirkendall)

Appendix

Casualty Reports

First Texas Infantry Regiment
Eltham's Landing, May 7, 1862

Col. A. T. Rainey comdg.

Field and Staff – Killed: Lt. Col. H. H. Black.

Company A, Lt. J. Waterhouse comdg. – Killed: Privates J. W. Etly, Thos. Mahon, T. Setzer, H. H. Hinnant. Wounded: Lt. W. W. Laney; Privates, Geo. L. Rogers, Pat Higgins, Hugh Hennesy. In action: Officers, 3; men, 40.

Company B, Lt. R. J. Harding comd. – Ina action: Officers 2; men, 13.

Company C, Capt. B. F. Perry comdg. – Killed: Lt. H. E. Decatur. Wounded: Corpl. R. B. Donnolly; Privates J. W. Trotter, Jo Taylor. In action: Officers, 3; men, 47.

Company D, Capt. W. M. Hewitt comdg. – Killed: Private C. F. Coy. Wounded: Corpl: J. F. McDowell, Private J. W. Smith. In action: Officers, 3; men, 60.

Company E, Capt. F. S. Bass comdg. – In action: Officers 3; men 40.

Company F, Capt. P. A. Work comdg. Killed: Private James Bush. Wounded: Sergt. R. B. May, Private E. T. Steedman. In action: Officers 3; men 32.

Company G, Lt. E. S. Jameson comdg. – Killed: Prvt. Martin O'Brien. Wounded: Privates R. C. McKnight, M. A. Knox. In action: Officers, 4; men, 30.

Company H, Capt. W. H. Gaston comdg. – Killed: Lt. John L. Spencer; Privates: S. B. Cornwell, W. A. Hosea, D. J. Hill, P. W. Mills. Wounded: Privates H. L. Martin, J. J. Foster. In action: Officers, 4; men, 66.

Company I, Capt. R. W. Comdg. – In action: Officers, 3; men, 32.

Company K, Capt. B. F. Benton comdg. Wounded; Private Jos. Lane. In action: Officers, 2: men, 40.

Company L, Capt. A. C. McKeen, comdg. – Killed: Privates Jos. F. Brown, Chas. L. Schadt. Wounded: Privates Frank Nichols, S. D. Simms, John Coffee. In action: Officers, 4; men, 71.

Total – Killed 15. Wounded 19. In action: Officers, 33; men, 471.[1]

Fourth Texas Infantry Regiment
Eltham's Landing, May 7, 1862

Col. John Marshall, comdg.

Field and Staff – In action: Officers, 2; men, 4.

Company A, Lt. S. H. Darden comdg. – In action: Officers, 3; men, 55.

Company B, Capt. B. F. Carter comdg. – In action: Officers 3; men, 78.

Company C, Capt. W. P. Townsend comdg. – In action: Officers, 4; men, 61.

Company D, Capt. John P. Bane comdg. – In action: Officers, 3; men, 50.

Company E, Capt. E. D. Ryan comdg. – In action: Officers, 2; men, 70.

Company F, Capt. Ed Cunningham comdg. – In action: Officers, 4; men, 50.

Company G, Capt. J. W. Hutcheson, comdg. – In action: Officers, 2; men, 80.

Company I, Lt. J. W. Lockridge comdg. – In action: Officers, 3; men, 37.

Company K, Capt. Wm. Martin comdg. – In action: Officers, 3; men, 50.

Total – Killed, 1. Wounded, 1. In action: Officers, 32; men, 581.[2]

---

1. Simpson, Harold B. Robertson, Jerome B. (eds,) *Touched With Valor: Civil War Papers and Casualty Reports of Hood's Texas Brigade Written and Collected by General Jerome B. Robertson, Commander of Hood's Texas Brigade 1862-1864*, Hillsboro, TX: Hill Junior College Press, 1964, 67-68.
2. Ibid, 79-80.

Fifth Texas Infantry Regiment
Eltham's Landing, May 7, 1862

Col. J. J. Archer comdg.

Company A, Capt. Farmer comdg. – In action: Officers, 3; men, 52.

Company B, Capt. Upton comdg. – Killed: Private Riley Scherer. Wounded: Privates C. Coffee, August Enke, Henry Senne, Hunt Terrell. Missing: W. J. Darden, F. K. Harris. In action: Officers, 3; men, 75.

Company C, Capt. Whaley comdg. – In action: Officers, 4; men 90.

Company D, Capt. Powell comdg. – In action: Officers 4, men, 55.

Company E, Capt. J. D. Rogers comdg. – In action: Officers, 2; men 58.

Company F, Capt. Bryan comdg. – In action: Officers, 1; men, 70.

Company G, Capt. J. C. Rogers comdg. – Wounded: Private A. J. Tomlinson. In action: Officers, 3; men, 49.

Company H, Lt. Robinson comdg. – In action: Officers, 3; men, 43.

Company I, Capt. Clay comdg. – In action: Officers 2; men 35.

Company K, Capt. Turner comdg. In action: Officers 3; men, 33.

Total – Killed, 2; wounded, 5; missing, 2. In action: Officers, 28; men, 562.[3]

GALVESTON DAILY NEWS
DECEMBER 19, 1891

Mr. A. M. Dillon, awoke some recollections of the Lone Star Rifles, and although I belonged to company E of the Fourth regiment, I have some memoranda pertaining to the Galveston company, among which are the casualties in the first regiments of the war and are as follows:

Eltham's Landing – Killed: James Brown, C Schodt. Wounded: Smith Sims, F. Nichols. J. Coffee.

Seven Pines – Wounded: J. W. Brown, W. A. Skelton.

---

3. Ibid, 85-86.

Gaines' Mill – Killed: Corporal J. L. Townsend, J Paupart. Wounded: James Nagle, S. D. Smith, H. Shulty, G. Hawkins, R. Jacobef.

Malvern Hill – Wounded: Captain W. A. Bedell, Corporal R. S. Robinson.

Bull Run – Wounded: Lieutenant J. M. Baldwin, Sergeant W. P. Randall, E. C. Corquodall.

Sharpsburg (Antietam) – Killed: Lieutenant J. C. S. Tomson, J. Frank. Wounded: Captain W. A. Bedell, Sergeant, Sergeant S. A. Carpenter, Corporal W. Zimmer, R. Jacobef, J. Hanson, J. T. Blessing, H. Cohen, P. Gillis, W. Hoskins, A. Jones, C. B. Halleck, C. H. Kingsley, J. Roake, F. Scharting, J. M. Smith, H. Schultz, J. Albrook, W. Leach, James Nagle, W. Young, - Welch.

I regret not having the list of all the battles, but these will perhaps recall to Galvestonians some familiar names of long ago.

John N. Henderson

BIBLIOGRAPHY

Books

Chilton, Frank, *Unveiling and dedication of monument to Hood's Texas brigade on the capitol grounds at Austin, Texas, Thursday, October twenty-seven, nineteen hundred and ten, and minutes of the thirty-ninth annual reunion of Hood's Texas brigade association held in Senate chamber at Austin, Texas, October twenty-six and twenty-seven, nineteen hundred and ten, together with a short monument and brigade association history and Confederate scrap book*, Houston, TX: F. B. Chilton, 1911.

Chesnut, Mary, *Diary of Mary Chesnut*, Fairfax, VA: Appleton & Company, 1905.

Davis, Nicholas A., *The Campaign From Texas To Richmond, With The Battle Of Fredericksburg*, Richmond, VA: Office of the Presbyterian Committee Of Publication of the Confederate States, 1863.

Fletcher, William, *Rebel Private, Front and Rear. Experiences And Observations From The Early Fifties Through The Civil War*, Beaumont, TX: Press Of The Green Print, 1908.

Folsom, J., *Heroes and Martyrs of Georgia. Georgia's record in the revolution of 1861*, Macon, GA: Burke, Boykin and Company, 1864.

Glover, Robert L., ed., *Tyler to Sharpsburg: The War Letters of Robert H. and William H. Gaston Company H, First Texas Infantry Regiment, Hood's Texas Brigade*, Waco, TX: Texian Press, 1960.

Hanks, O. T., *History Of Captain B., F. Benton's Company, Hood's Texas Brigade, 1861-1865*, Waco, TX: Harrison Books, 1984.

Hood, John B., *Advance and Retreat: Personal Experiences in the United States and Confederate Armies*, New Orleans, LA: Beauregard, 1880.

Laswell, Mary.(ed.), *Rags and Hope, The Memoirs of Val C. Giles, Four Years with Hood's Brigade, Fourth Texas Infantry, 1861-1865*, Coward-McCann: New York, 1961.

Lemon, Mark, *Feed Them The Steel! Being, the Wartime Recollections of Capt. James Lile Lemon, Co A, 18th Georgia Infantry CSA*, Self-published, 2013.

G. Otott, G., *"Clash in the Cornfield: The 1st Texas Volunteer Infantry in the Maryland Campaign," Antietam the Maryland Campaign of 1862*, Campbell, CA: Savas Publishing Company, 1997.

Owen, Joseph L. and Drais, Randy L., *Texans at Gettysburg: Blood and Glory with Hood's Texas Brigade*, Charleston, SC: Fonthill Media, 2016.

Palmer, William A. Jr., *The Battle of Eltham's Landing, May 7, 1862*, Self-Published, 2012.

Polley, J. B., *A Soldiers Letter to Charming Nellie*, New York, NY: The Neale

Publishing Company, 1908.

_____, *Hood's Texas Brigade, Its Marches, Its Battles, Its Achievements*, New York, NY: The Neale Publishing Company, 1910.

Robertson, Jerome B. Simpson, Harold B. (eds), *Touched With Valor: Civil War Papers and Casualty Reports of Hood's Texas Brigade Written and Collected by General Jerome B. Robertson, Commander of Hood's Texas Brigade 1862-1864*, Hillsboro, TX: Hill Junior College Press, 1964.

Schadt, Stuart E., *Letters Home Charles and William Schadt, 1861-1865*, Bradley Stuart Books, 2015.

Schmutz, J., *"The Bloody Fifth." The 5th Texas Infantry Regiment, Hood's Texas Brigade, Army of Northern Virginia, Vol. 1: Secession to the Suffolk Campaign*. El Dorado Hills, CA: Savas Beatie, 2016.

Simpson, Harold B., *Hood's Texas Brigade: A Compendium*, Hillsboro, TX: Hill Junior College Press, 1977.

_____ *Hood's Texas Brigade: Lee's Grenadier Guard*, Hillsboro, TX: Hill Jr. College Press, 1970.

_____ *Gaines' Mill to Appomattox, Waco & McLennan County In Hood's Texas Brigade*, Waco, TX: Texian Press, 1963.

Skoch, George and Perkins, Mark W., (eds), *Lone Star Confederate: A Gallant and Good Soldier of the Fifth Texas Infantry*, (College Station, TX: Texas A&M University Press, 2003.

The War of the Rebellion: A Compilation of the Official Records of the Union and Confederate Armies, 128 Vols. (Washington DC, 1880-1901), Series 1.

Todd, G., *First Texas Regiment*, Waco, TX: Texian Press, 1964.

Warner, Ezra J., *Generals in Blue: Lives of the Union Commanders*, Baton Rouge, LA: Louisiana State University Press, 1964.

Williams, Edward, *Hood's Texas Brigade in the Civil War*, Jefferson, NC: McFarland & Company Inc., 2012.

Yeary, Mamie, *Reminiscences of the boys in gray, 1861-1865*, Dallas: Wilkinson Printing Company, 1912.

Manuscripts/Letters

Judy Callaway Ostler Family Letters/Manuscripts

Col. Alexis Theodore Rainey: First Texas Regiment, personal non-published compilation.

A. T. Rainey to Anne Rainey, August 28, 1861, in Rainey Family Papers, Judy Callaway Ostler.

Texas Heritage Museum – Historical Research Center.

Letter, William L. "W. L." Edwards to Roxie Edwards, May 13, 1862. Hood Texas Brigade files. Texas Heritage Museum-Historical Research Center.

Letter, W. L. Edwards to Travis and Elizabeth Scott, May 10, 1862. Hood's Texas Brigade files, Texas Heritage Museum-Historical Research Center.

Letter, W. L. Edwards to Roxie Edwards, May 13, 1862, Hood's Texas Brigade files, Texas Heritage Museum-Historical Research Center.

Letter, Samuel T. Owen to Mother, May 22, 1862, Hood's Texas Brigade files. Texas Heritage Museum-Historical Research Center.

J. B. Polley, Diary, Unpublished manuscript, Hood's Texas Brigade files, Texas Heritage Museum-Historical Research Center.

John H. Reagan to Anne Rainey, May 9, 1862, in Rainey Family Papers, Judy Callaway Ostler.

Leila Reeves Eads, Malachiah Reeves, Memoirs of the mercies of a covenant God, while traveling through the wilderness of this world of life: being the autobiography of Malachiah Reeves, originally from middle Alabama and through much of East Texas, and now of Ranger, middle West Texas, Unpublished manuscript, Hood's Texas Brigade files, Texas Heritage Museum.

William Townsend to unknown, May 11, 1862, Hood's Texas Brigade Files, Texas Heritage Museum.

Mark S. Womack to unknown, circa May 1862, Hood's Texas Brigade files, Texas Heritage Museum-Historical Research Center.

Magazines/Periodicals

Anderson County Historical Society. *Anderson County In The Civil War, in The Tracings*, (Palestine, TX, 1986), Volume 05, Number 01, Winter 1986.

B. L. Aycock, "The Lone Star Rifles," in *Confederate Veteran*, February 1923, Volume 30. No. 2.

Chuck Haas, *Hood's Texans – "An irresistible Attacking Force,"* in *The Junior Historian* (December 1965), vol. 26.

J. T. Hunter, "At Yorktown In 1862 And What Followed" in *Confederate Veteran*, (1918), Vol. 26.

William A. Palmer Jr., "Good Bye Dear Sister," The Unhappy War Of A Texan Who Was The Last To Enlist And The First To Fall," in *Civil War Times*, October 2016.

Tommy Wilson, "The Last Will of Matt Dale," in *The Junior Historian*, (May 1955).Volume 16, Number 6.

Newspapers

Austin (TX) State Telegraph

Austin (TX) Weekly Statesman

Bryan (TX) The Eagle

Charlotte (NC) Evening Bulletin

Clarksville (TN) Weekly Chronicle

Fayetteville (NC) Weekly Observer

Galveston (TX) Daily News

Galveston (TX) Weekly News
Houston (TX) Telegraph
Halletsville (TX) Herald
Houston (TX) Tri-Weekly Telegraph
Houston (TX) Weekly Telegraph
Marshall (TX) Messenger
Natchez (MS) Weekly Courier
New Orleans (LA) Times-Democrat
Raleigh (NC) Register
Raleigh (NC) The Spirit of the Age
Richmond (VA) Dispatch
Richmond (VA) Observer
San Antonio (TX) Daily Express
San Antonio (TX) Express
St. Albans (VT) Weekly Messenger
Temple (TX) Daily Telegram
Temple (TX) Eagle
Windsor (VT) Vermont Journal
Winston-Salem (NC) Western Sentinel

Web Sources

Handbook of Texas Online, Jennifer Bridges, "Black, Harvey H.," accessed April 22, 2018, http://www.tshaonline.org/handbook/online/articles/fb174.

Arthur H. Edey, accessed October 16, 2018, https://siegelauctions.com/lot_grd.php?majgroup=United%20States&cat_supgroup=Confederate%20States%20and%20Civil%20War-Related&recsperpage=10&lot_catfk=104&subgroup=&realized1=&realized2=&sale_no=&srtorder=Edey&lot_no=&sdate1=01%2F01%2F1930&sdate2=01%2F01%2F2020&symbol[]=All&lotclass=All&syear=All&pfoper=All&pseoper=All&pfgrade=&psegrade=&gandor=or&keyword=&catselect=eq&pscolumn=default&pssortby=&sortord=DESC&photo=&calledfrom=lkp.

Handbook of Texas Online, Diana J. Kleiner, "William A. Fletcher," accessed October 16, 2018, http://www.tshaonline.org/handbook/online/articles/ffldd.

Vicki Ragan Harris, "William P. Townsend." Accessed April 12, 2018, https://www.findagrave.com/memorial/145239127/william-purnell-townsend.

Jones, Dan, "Bennett Wood," Accessed: April 8, 2018. https:222.findagrave.com/memorial/75741835/bennett-wood.

Joe Joskins, *A Sketch of Hoods Texas Brigade of the Virginia Army by Joe Joskins A rebel in Co "A" 5th Texas Vols "Hoods Texas Brigade" "Fields Division" "Longstreet's Corps," Army of Northern Virginia, June 18, 1865*, UTSA Digital Libraries, Rare books collection, July 18, 2012 Accessed April 8, 2018. http://digital.utsa.edu/cdm/compoundobject/collection/p15125coll10/id/8440/rec/1.

Handbook of Texas Online, Jennifer Bridges, "Black, Harvey H.," accessed April 22, 2018, http://www.tshaonline.org/handbook/online/articles/fb174.

Michelle, Alfred J. Wilson, Accessed April 10, 2018. https:www.findagrave.com/memorial/791472/Alfred-j-wilson.

Handbook of Texas Online, Donald F. Reynolds, "Marshall, John F.," accessed September 28, 2017, http:www.tshaonline.org/handbook/online/articles/fma35.

James T. Hunter, Granbury Texas Brigade Website, accessed March 27, 2018, http://www.granburystexasbrigade.org/gallery/heroes/officers/hunter_james.html.

Handbook of Texas Online, "Sayers, Joseph Draper," accessed April 22, 2018, http://www.tshaonline.org/handbook/online/articles/fsa41.

Gail Ray, "O. T. Hanks," accessed July 29, 2018. https://www.findagrave.com/memorial/49891967/orlando-thacker-hanks.

Larry Scarbourough "Mark Sanders Womack," Accessed May 20, 2018. https://www.findagrave.com/memorial/26941693/mark-sanders-womack/photo.

Index

About the Author

Joe Owen is a National Park Ranger at Lyndon B. Johnson National Historical Park in Johnson City, Texas. Having served 15 years in the U.S, Navy, he attended college at East Central University in Ada, Oklahoma receiving a Bachelor's Degree in History and a Master's Degree in Secondary Education. He taught Social Studies for eight years in Oklahoma, Texas and Oregon before working for the National Park Service.

He is a co-author of two books about Hood's Texas Brigade, Texans At Gettysburg: Blood and Glory with Hood's Texas Brigade, (2016), Texans at Antietam: A Terrible Clash of Arms, September 16-17, 1862, (2017), and is the author of Lone Star Valor: Texans of the Blue and Gray at Gettysburg (2019.) He received the Jefferson Davis Gold Medal from the United Daughters of the Confederacy for outstanding research and writing in 2019.

Joe lives in Blanco, Texas.

www.ingramcontent.com/pod-product-compliance
Ingram Content Group UK Ltd.
Pitfield, Milton Keynes, MK11 3LW, UK
UKHW020143250726
13967UKWH00002B/842

9 781945 602207